Innovation, Technology, and Knowledge Management

Series Editor
Elias G. Carayannis, George Washington University, Washington D.C., USA

For further volumes:
http://www.springer.com/series/8124

Steven P. MacGregor • Tamara Carleton
Editors

Sustaining Innovation

Collaboration Models for a Complex World

Springer

Editors
Steven P. MacGregor
IESE Business School
Barcelona, Spain
smacgregor@iese.edu

Tamara Carleton
Innovation Leadership Board LLC
Silicon Valley, USA
carleton@innovation.io

ISBN 978-1-4614-2076-7 e-ISBN 978-1-4614-2077-4
DOI 10.1007/978-1-4614-2077-4
Springer New York Dordrecht Heidelberg London

Library of Congress Control Number: 2011943346

Printed on acid-free paper

Springer is part of Springer Science+Business Media (www.springer.com)

Series Foreword

The Springer book series *Innovation, Technology, and Knowledge Management* was launched in March 2008 as a forum and intellectual, scholarly "podium" for global/local, transdisciplinary, transsectoral, public–private, and leading-/"bleeding"-edge ideas, theories, and perspectives on these topics.

The book series is accompanied by the Springer *Journal of the Knowledge Economy*, which was launched in 2009 with the same editorial leadership.

The series showcases provocative views that diverge from the current "conventional wisdom," which are properly grounded in theory and practice, and that consider the concepts of *robust competitiveness*,[1] *sustainable entrepreneurship*,[2] and *democratic capitalism*,[3] central to its philosophy and objectives. More specifically, the aim of this series is to highlight emerging research and practice at the dynamic intersection of these fields, where individuals, organizations, industries, regions, and nations are harnessing creativity and invention to achieve and sustain growth.

Books that are part of the series explore the impact of innovation at the "macro" (economies, markets), "meso" (industries, firms), and "micro" levels (teams, individuals),

---

[1] We define *sustainable entrepreneurship* as the creation of viable, profitable, and scalable firms. Such firms engender the formation of self-replicating and mutually enhancing innovation networks and knowledge clusters (innovation ecosystems), leading toward robust competitiveness (E.G. Carayannis, *International Journal of Innovation and Regional Development* 1(3), 235–254, 2009).

[2] We understand *robust competitiveness* to be a state of economic being and becoming that avails systematic and defensible "unfair advantages" to the entities that are part of the economy. Such competitiveness is built on mutually complementary and reinforcing low-, medium-, and high-technology and public- and private-sector entities (government agencies, private firms, universities, and nongovernmental organizations) (E.G. Carayannis, *International Journal of Innovation and Regional Development* 1(3), 235–254, 2009).

[3] The concepts of *robust competitiveness* and *sustainable entrepreneurship* are pillars of a regime that we call "*democratic capitalism*" (as opposed to "popular or casino capitalism"), in which real opportunities for education and economic prosperity are available to all, especially—but not only—younger people. These are the direct derivatives of a collection of top-down policies as well as bottom-up initiatives (including strong research and development policies and funding, but going beyond these to include the development of innovation networks and knowledge clusters across regions and sectors) (E.G. Carayannis and A. Kaloudis, *Japan Economic Currents*, p. 6–10 January 2009).

drawing from such related disciplines as finance, organizational psychology, research and development, science policy, information systems, and strategy, with the underlying theme that for innovation to be useful it must involve the sharing and application of knowledge.

Some of the key anchoring concepts of the series are outlined in the figure below and the definitions that follow (all definitions are from E.G. Carayannis and D.F.J. Campbell, *International Journal of Technology Management*, 46, 3–4, 2009).

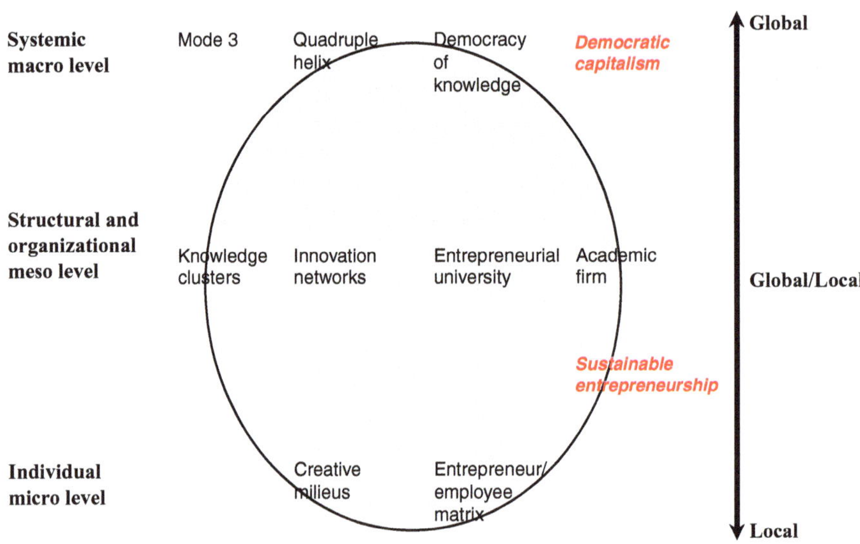

Conceptual profile of the series *Innovation, Technology, and Knowledge Management*

- The "Mode 3" Systems Approach for Knowledge Creation, Diffusion, and Use: "Mode 3" is a multilateral, multinodal, multimodal, and multilevel systems approach to the conceptualization, design, and management of real and virtual, "knowledge-stock" and "knowledge-flow," modalities that catalyze, accelerate, and support the creation, diffusion, sharing, absorption, and use of cospecialized knowledge assets. "Mode 3" is based on a system-theoretic perspective of socio-economic, political, technological, and cultural trends and conditions that shape the coevolution of knowledge with the "knowledge-based and knowledge-driven, global/local economy and society."
- Quadruple Helix: Quadruple helix, in this context, means to add to the triple helix of government, university, and industry a "fourth helix" that we identify as the "media-based and culture-based public." This fourth helix associates with "media," "creative industries," "culture," "values," "life styles," "art," and perhaps also the notion of the "creative class."

- Innovation Networks: Innovation networks are real and virtual infrastructures and infratechnologies that serve to nurture creativity, trigger invention, and catalyze innovation in a public and/or private domain context (for instance, government–university–industry public–private research and technology development coopetitive partnerships).
- Knowledge Clusters: Knowledge clusters are agglomerations of cospecialized, mutually complementary, and reinforcing knowledge assets in the form of "knowledge stocks" and "knowledge flows" that exhibit self-organizing, learning-driven, dynamically adaptive competences and trends in the context of an open systems perspective.
- Twenty-First Century Innovation Ecosystem: A twenty-first century innovation ecosystem is a multilevel, multimodal, multinodal, and multiagent system of systems. The constituent systems consist of innovation metanetworks (networks of innovation networks and knowledge clusters) and knowledge metaclusters (clusters of innovation networks and knowledge clusters) as building blocks and organized in a self-referential or chaotic fractal knowledge and innovation architecture[4], which in turn constitute agglomerations of human, social, intellectual, and financial capital stocks and flows as well as cultural and technological artifacts and modalities, continually coevolving, cospecializing, and cooperating. These innovation networks and knowledge clusters also form, reform, and dissolve within diverse institutional, political, technological, and socioeconomic domains, including government, university, industry, and nongovernmental organizations and involving information and communication technologies, biotechnologies, advanced materials, nanotechnologies, and next-generation energy technologies.

To whom is this book series directed? The book series addresses a diversity of audiences in different settings:

1. *Academic communities.* Academic communities worldwide represent a core group of readers. This follows from the theoretical/conceptual interest of the book series to influence academic discourses in the fields of knowledge, also carried by the claim of a certain saturation of academia with the current concepts and the postulate of a window of opportunity for new or at least additional concepts. Thus, it represents a key challenge for the series to exercise a certain impact on discourses in academia. In principle, all academic communities that are interested in knowledge (knowledge and innovation) could be tackled by the book series. The interdisciplinary (transdisciplinary) nature of the book series underscores that the scope of the book series is not limited a priori to a specific basket of disciplines. From a radical viewpoint, one could create the hypothesis that there is no discipline where knowledge is of no importance.
2. *Decision makers—private/academic entrepreneurs and public (governmental, subgovernmental) actors.* Two different groups of decision makers are being addressed simultaneously: (1) private entrepreneurs (firms, commercial firms,

[4] E.G. Carayannis, *Strategic Management of Technological Learning*, CRC Press, 2000.

academic firms) and academic entrepreneurs (universities), interested in optimizing knowledge management and in developing heterogeneously composed knowledge-based research networks; and (2) public (governmental, subgovernmental) actors that are interested in optimizing and further developing their policies and policy strategies that target knowledge and innovation. One purpose of public *knowledge and innovation policy* is to enhance the performance and competitiveness of advanced economies.

3. *Decision makers in general.* Decision makers are systematically being supplied with crucial information, for how to optimize knowledge-referring and knowledge-enhancing decision-making. The nature of this "crucial information" is conceptual as well as empirical (case-study-based). Empirical information highlights practical examples and points toward practical solutions (perhaps remedies), conceptual information offers the advantage of further-driving and further-carrying tools of understanding. Different groups of addressed decision makers could be decision makers in private firms and multinational corporations, responsible for the knowledge portfolio of companies; knowledge and knowledge management consultants; globalization experts, focusing on the internationalization of research and development, science and technology, and innovation; experts in university/business research networks; and political scientists, economists, and business professionals.
4. *Interested global readership.* Finally, the Springer book series addresses a whole global readership, composed of members who are generally interested in knowledge and innovation. The global readership could partially coincide with the communities as described above ("academic communities," "decision makers"), but could also refer to other constituencies and groups.

Elias G. Carayannis
Series Editor

Foreword

Sustainability means different things to different people. I do not care about the definition of sustainability, but I care about the future generations of this planet and how they can lead better lives. This care of mine, reflected throughout Japan, is inherited from our ancestors, who worked hard and hoped to make a better world. Their wishes still run in our bodies and minds, as well as our society and environments. Sustainability is built into our genes, and we will deliver this to our descendants.

As a part of my work, I am engaged in several activities at the National Institute of Science and Technology Policy in Japan. The Institute develops forecasts in science and technology and studies various topics that impact our society and life, including regional and global innovation systems, private and public intellectual communities, and educational systems for current and future generations. This task necessarily involves engagement with a wide variety of people from different backgrounds and spheres, each with their own understanding and needs on often complex subjects.

This complex collaboration is true also in my other work, which has crossed multiple institutional spheres. I have worked at private companies, both domestically and globally—such as Toshiba, IBM, and SCSK Corporation (formerly CSK)—for research and business. My time is also devoted to educational tasks at different academic institutions where I teach a wide range of subjects from information systems management to design thinking for innovation and sustainability.

This is why I am encouraged to see this volume, *Sustaining Innovation: Collaboration Models for a Complex World*, which offers a practical set of real-world examples, visions, and research insights for today's policymakers, as well as business leaders and academics.

Sustainability requires change. From a Japanese perspective, I have seen my nation overcome and enjoy many changes to its industrial base, economy, and demographics to become one of the most innovative countries of the last 50 years. Yet we face again new and complex challenges, such as the recent earthquake and tsunami. The Japanese islands, standing at the edge of the Eurasian continent facing the Pacific Ocean and the Americas beyond, have brought varying degrees of change; people, things, and ideas come from both sides of the world, east and west. And now, ever-advancing technology accelerates that change to drive innovation.

So, for those of you engaged in the pursuit of sustaining innovation and building a better future for society, no matter if you approach the opportunity from a policy, academic, or industry domain, I commend this volume, edited by Steven MacGregor and Tamara Carleton, as a means of helping you design and deliver that future. It will certainly be one of the resources I keep on my bookshelf.

Tokyo, Japan Toshiaki Kurokawa

Introduction

Evolutionary scientist Charles Darwin once noted, "In the long history of humankind (and animal kind, too) those who learned to collaborate and improvise most effectively have prevailed." In many ways, the process of innovation is a constant social dance, where the best dancers thrive by adapting new steps with multiple partners. The systematic and continuous generation of value in any innovation system relies on collaboration between different groups, who must overcome multiple, often competing, agendas and needs to work together fruitfully over the long term. In this collection of essays and viewpoints, we investigate different combinations of collaborative relationships between innovation actors, many of which are changing conventional expectations of institutional relationships.

In short, no particular combination has emerged as the most dominant, or even resilient, model of innovation. Several of the authors in this volume expand on our understanding of the triple helix model, with both academics and practitioners looking to the quadruple helix as the new standard. Other authors address aspects of open innovation, co-creation, user-centered design, and mass customization—all testaments to the rapidly shifting landscape. At the same time, many businesses, academics, and governments, not to mention nonprofit organizations, foundations, and society at large, are active in conversations about how to pursue a more sustainable model of innovation. The pursuit of this holy grail of innovation is both facilitated and complicated by an ever-accelerating technological landscape in which social networking and mobile tools are emerging as new dance arenas.

On Sustainability

First, a disclaimer about the book's central theme: we know that the word sustainability is quickly following that of innovation into buzz-word hell that makes it almost meaningless in certain situations, or at least invisible due to its omnipresence. Yet sustainability is wholly appropriate for our context for two main reasons. First, the economic crisis of recent years makes us reflect on the durability of any success we may enjoy, and second, we investigate the role of society in collaboration for that durability. We therefore consider sustainability from the *value* perspective—the enduring competitive position of an organization and its ability to innovate on a continual basis as shown by its market results—as well as the *values* perspective in which the society-wide context of sustainable development is complemented by the role of business within society through corporate social responsibility (CSR). The gerund of sustainability (in other words, the "-ing") is also used on purpose, since the word emphasizes the work in progress of continuous innovation, as well as the action, dynamism, and scale of the challenge.

Value and values have rarely been comfortable bedfellows, at times due to problems regarding the nuances of language and resulting misconceptions, yet there is much common ground to be exploited in this increasingly collaborative era. Both value and values, as we describe them here, necessarily take note of a broad church of stakeholders, be they end users, suppliers, employees, or other actors in an innovation ecosystem.

A growing number of scholars and practitioners have addressed the links between the two concepts. Michael Porter's thesis on shared value (Porter and Kramer 2011) attempts to move notions of responsibility and fairness closer to the economic mainstream by linking values with value, and, in spite of some criticism, is a positive contribution to the field of economic management. Porter advises on dropping the CSR term, partly given the US-centric view that CSR is closely associated to philanthropy, yet Europe has traditionally taken a broader view on the term with, for example, previous work on Corporate Social *Opportunity* (Grayson and Hodges 2004) helping to influence Porter's current thesis.

In our view, CSR, by whatever name, is important to an understanding of sustainability because it provides a set of guidelines and mission-driven ethos that there ought to be a better way of doing things that is in the best interests of all in the long term. Innovators, including William McDonough with the Cradle-to-Cradle concept, show a possible future vision that brings together elements of both value and values. In other words, what an organization believes (values) can powerfully influence what is produced (value). What is this smarter way of working called? We challenged our authors to consider their own definitions and terminology.

Ultimately, we believe that the systematic, continuous generation of value can only be achieved when all stakeholders are considered and share in that value or, at the very least, are not victims of others' value gain. As an example, consider the rapid growth of Silicon Valley, which continues to dominate innovation rankings. The region exemplifies broad-based employee ownership, in which nonexecutive employees and line workers have the opportunity to own company shares and stock options in their own companies, allowing them to benefit from their knowledge and labor alongside management and investors. Practices in good corporate citizenship and user-driven innovation have similar visions of broad stakeholder involvement at their core, engaging extensive communities of users in the process of value creation. These examples and more confirm to us that a multilayered consideration of sustainability, in which responsibility, inclusiveness, and competitiveness are intertwined, is both necessary and meaningful to our investigation of complex collaboration models.

On Global Networks

Global interactions add another dimension to the complexity. Collaboration with a spectrum of external partners—including foreign competition, international alliances and consortia, and technical communities—has growing increasingly commonplace in the last few decades. Innovation network experts Walter Powell and Stine Grodal (2005) find that, "Complex networks of firms, universities, and government labs are critical features of many industries, especially so in fields with rapid technological progress, such as computers, semiconductors, pharmaceuticals, and biotechnology" (pg. 58). Social media and other online tools have accelerated the uptake of distributed networks of practice, whose users are often spread over location and international research and development (R&D) sites.

However, it is easy to get carried away with the recent explosion in social media, and many have subscribed to the fact of a completely flat world (Friedman 2005). Yet the data shows that, far from being completely global, local differences do still matter (Ghemawat 2011). Only 2% of students are at universities outside their home countries, and only 3% of people live outside their country of birth. In addition, less than 1% of all American companies have any foreign operations, while Foreign Direct Investment (FDI) accounts for only 9% of all fixed investment. These numbers help us to reset our assumptions about technology's actual impact on global collaboration.

We consider, therefore, that the difficulty in sustaining innovation could be due to a lack of understanding of local differences and specific needs of regional stakeholders. Our aim is to uncover enduring models of smarter work by learning how different regions have redefined and adapted practices of collaboration in a globally networked world. We deliberately asked the authors to take a global multistakholder perspective, and through their stories, they describe and analyze established innovation hotspots in the USA and Europe, as well as emerging and rapidly growing economies, including India and Africa. By finding new examples from around the world, we hope to expand on the usual discussions of collaboration and add different voices to the overall dialogue.

Models of Complex Collaboration

The state of knowledge about innovation collaboration continues to grow. The classic model is the triple helix, which refers to the innovation triad of business, academia, and government. Introduced by European professor Henry Etzkowitz (1996) some 15 years ago, this simple model helps explain the dynamic interplay occurring between these three institutional spheres. In their classic roles, government authorities provide the social mission, regulation requirements, and funding; academic groups lead in research and discovery efforts; and businesses find new market opportunities that put the resulting innovation into the hands of users. Etzkowitz posited that these roles were becoming increasingly shared and interdependent.

Businesses, universities, and governments—previously relatively separate and distinct institutions—were now assuming tasks in the development of new technologies that were once the province of the other.

During the late 1990s and 2000s, the mainstream adoption of the Internet influenced practices of distributed working and applications of social computing. The individual became involved in broad-scale innovation in more ways, and power dynamics shifted slightly in the original model of the triple helix. Collaboration became more fluid within and between the different institutional spheres, and the notion of a quadruple helix emerged that blended in the perspective of a media-based and culture-based public (Carayannis and Campbell 2009). The new fourth axis encompasses civil society in terms of consumers, nongovernmental organizations, and citizens—many of whom are active in designing their own services, providing input at key points in the innovation process, and determining the final value of innovation output. A more recent notion is the quintuple helix, which frames knowledge and innovation in the context of the natural environment with an overarching goal of sustainable development (Carayannis and Campbell 2010).

Other models have gained traction in innovation circles, particularly with corporate management. The philosophy of user-centered design through books such as *The Design of Everyday Things* (Norman 1986) puts the needs of users at the center of the standard product design and development process. Many see consumer electronics manufacturer Apple at the forefront of this movement. Another model of open innovation, first described by Henry Chesbrough (2003), addresses collaboration in terms of outsourcing partners, challenging notions of inventing from within an organization. In this spirit, consumer products company Procter & Gamble transformed their R&D process several years ago, requiring internal product teams to source at least 50% of their ideas for new or refined products from outside the company (Huston and Sakkab 2006).

Governments have also experimented with different models of collaboration. The early 1990s saw a rise in the concept of National Innovation Systems, described by the OECD (1997) as "the network of institutions in the public and private sectors whose activities and interactions initiate, import, modify, and diffuse new technologies." OECD observed that previous analysis focused primarily on innovation inputs (such as research expenditures) and outputs (such as patents). In knowledge-based economies, the interactions between actors involved in technology development are equally as important as investments in R&D because the actors are key to translating the inputs into outputs. An active flow of knowledge, such as the movement of personnel or joint industry research, ensures the smooth operation of national innovation systems.

Many more flavors of collaboration abound. The truth is, the dialogue is still emerging, and as the world evolves, more groups will continue working together in new and different ways. Through the collected stories in this volume, we aim to take another step forward in understanding and defining what complex collaboration is about in the pursuit of innovation.

Multiple Viewpoints on Sustaining Innovation

Toward that end, we invited a range of authors from different regions, industries, and professions to describe their work firsthand with complex collaboration. We asked them: What new models have they been building and experiencing? What lessons could they share with others interested in sustainable innovation? How did they engage different groups and ensure both value- and values-driven interaction within and across their innovation networks and communities?

As we heard from our colleagues around the world, we encountered an unexpected challenge in organizing their stories. Our first impulse was to organize stories by the lead actor, the institutional sphere responsible most for driving change. Many of the examples involve the consumer or civil society in some fashion; all the stories cross the three institutional spheres of influence of the triple helix. However, we felt that this approach perpetuated institutional silos, and many of the authors described activities beyond a group's usual role.

Then we considered presenting the stories in terms of theory and practice. We found many of the professors studied and even intervened in their industry cases, and our contributing practitioners reflected deeply on the abstract principles underlying action. Next, we debated presenting the stories in terms of their regional context. Then again, we found many examples crossed geographic borders.

Finally, we identified a more provocative and meaningful structure that revealed the state of development in each author's collaboration model: Visions, Research, and Experiences. Some models are in the formative stage of inspiration (Visions), others are under analysis and critical interpretation (Research), and a third set are in deep experimentation and implementation mode (Experiences). These three sections of Visions, Research, and Experiences are each comprised of several chapters, all of which are described and connected below.

Section I: Visions

In the first section, the authors present their visions of collaboration. They imagine future possibilities that are realistic, credible, and attractive to all the organizations involved. While some pieces exist today, the bulk of effort, alignment, and integration must happen over time. These visions orient everyone to future possibility and provide positive direction or even a call to arms for a new model of collaboration.

In Chap. 1, Nam Mokwunye, a visiting scholar at Stanford University in the USA, offers a broad and ambitious vision on sustaining innovation in the African continent. He focuses on the telecommunication sector, which is heavily regulated, fragmented, and disorganized with inadequate infrastructure in multiple African economies. Clarifying the quadruple helix model, Mokwunye distinguishes between nongovernment groups in civil society and end users in their influence on practices of local technology transfer. With a different twist on innovation networks, his vision unifies physical, social, and virtual networks in the pursuit of scalable and sustainable innovation in African telecom.

Laura Ahonen and Tuija Hämäläinen from the City of Jyväskylä in Finland report on the groundbreaking CLIQ project in Chap. 2. CLIQ attempted to gain a practical, policy-based view on the quadruple helix model and was conducted by public administrations in small- and medium-sized European cities. The authors report on the resultant visions for the practical implementation of multihelix collaboration, including a benchmarking methodology, blueprint, and toolkit for innovation. They present their learning journey in discovering the notion of the quadruple helix and how it can be supported.

In Chap. 3, Josep Lluis de la Rosa and Andrea Bikfalvi of the University of Girona in Spain describe their vision for reinventing the practice of carpooling for the digital age, with a spin-off company currently putting their vision into action. Through the creation of the First Bank of Cents, they show how social currencies, web 2.0, and GPS technology can influence good citizen behavior and corporate responsibility around a common cause. Their bold idea has precedence, building on other forms of social commuting—such as the successful American practice of casual carpooling, known as "slugging" in the Washington, DC area, that has been underway since 1975.

Section II: Research

The second section presents a set of research projects with the authors taking a critical view of the causes and developments in a particular context, including the Swedish healthcare system, the Indian design industry, and European textiles. The first two chapters focus on actor collaboration in the triple helix while a case study approach is taken in the final two chapters.

In Chap. 4, Andreas Larsson, Susanna Bill, Jenny Ingridsdotter, and Annika Olsson address the changes underfoot in the Swedish healthcare system, particularly related to the design and procurement of new Swedish healthcare devices. Drawing from their experiences as Swedish research scholars over a 3-year project, the authors describe some of the lessons and challenges that the different project groups encountered during their regular learning sessions.

From the Ideas Lab at the Indian Institute of Science in India, Chap. 5 shows Gokula Vijaykumar A.V. and Amaresh Chakrabarti's report on the findings of two Indian companies in the engineering design industry. They focus on the critical role of knowledge processes in industrial design collaboration. Finding current knowledge process models to be inadequate to support design collaboration, they propose the KRIT model and a related Influence model, which show how the elements of Knowledge, Requirements, Interactions, and Tasks may better support design work.

Alexandra Simon and Pilar Marquès, also from the University of Girona in Spain, investigate how fourth-pillar organizations can genuinely support triple helix collaboration in Chap. 6. Through four case studies in three European countries, they develop the concept of these hybrid organizations and how their use in Europe differs from that in the USA and Canada, offering in the process an alternative European model to complement previously published work on the subject.

In Chap. 7, Nigel Roome and Céline Louche from Vlerick Business School in Belgium take a deeper look at Rohner Textil AG, a Swiss textile company that, in collaboration with William McDonough, was one of the first ever companies to apply the Cradle-to-Cradle concept on a business level. They describe how Rohner updated its corporate vision in partnership with DesignTex, an American fabric design company. Rohner then engaged other groups in its production network—including wool farmers, chemical dye suppliers, loom equipment providers, and government regulators—to realize an ambitious vision of quality compostable products and process management. By studying Rohner, the authors demonstrate that genuine sustainable development requires shared leadership and active alignment of multiple stakeholders. They also show that learning at one company can influence practices and applications in other networks globally.

Section III: Experiences

The final section of Experiences emphasizes networks of complex collaboration in action. Three of the chapters reflect on corporate-driven networks that have brought a diverse mix of innovation partners and stakeholders together to find new market opportunities and enhance the user experience. The other chapter describes a recent government-led effort that has redefined regional industry relationships and added a layer of virtual interactions to the physical networking. All authors pursued an interest in long-term value and network sustainability.

Álvaro Morón Alonso, innovation manager at the Spanish bank BBVA, together with Capgemini consultant Javier Sebastián Cermeño detail the innovation network that BBVA has constructed to support their innovation endeavors in Chap. 8. They illustrate a fascinating case of how a financial services firm can have a broad interest in innovation outside of their core business, sharing insider details about the current global configuration of the network and IDEO's role in the design of a new generation of cash machines.

Chap. 9 presents the experiences of innovation expert David Coates, who has served an influential role in building and enhancing the knowledge networks served by the Technology Strategy Board, the national innovation agency of the UK. The agency's goal is to accelerate economic growth by stimulating and supporting business-led innovation, partly by bringing different organizations together in a complex web of business, academic, and government communities. Coates describes a new online platform that facilitates knowledge sharing for the country's broad network of small businesses, and he explains the unique role a community manager must serve to ensure a vibrant and self-sustaining exchange within the network.

In Chap. 10, Dagmar Chlosta explains how she introduced the use of virtual samples in the product development and production process at a global sporting goods company. Adopting techniques from 3D visualization, Chlosta led a transformation process similar to how Boeing designed its 777 aircraft entirely by computer during the early 1990s. The process at the sporting goods company required changing the mind-set of all the groups involved, and Chlosta encouraged a shift in values through multiple tactics, including training partners and engaging customers at key points.

Bringing the Experiences section and the book to a close in Chap. 11, New Zealander futurist Roger Dennis, British strategist Tim Jones, and British executive Leo Roodhart look to the future. Opening with the influential case of Shell's

GameChanger initiative of the early 2000s, they recount the development of foresight science, focusing on Vodafone's Future Agenda program as an example of an open source network. Recognizing the importance of a global outlook balanced by local differences, the Future Agenda team held 50 workshops in 25 locations around the world, gathering insights from over 2,000 people.

In sum, we have attempted to encourage, guide, and ultimately represent a global conversation in a way that makes you, as the reader, feel compelled to join. This conversation is made up of 22 authors from Belgium, Finland, Germany, India, New Zealand, Nigeria, Spain, Sweden, UK, and the US, representing the full scope of the triple helix and beyond. We have attempted to balance business cases (Chaps. 7 and 10) with bold visions (Chap. 1), insightful research that focuses on key units of analyses such as collaboration (Chaps. 4 and 5) with practical-based research based on key projects (Chaps. 2 and 6), as well as entrepreneurial endeavors that such projects aim to support (Chap. 3). Finally, we have shown large-scale networks in action (Chaps. 8, 9, and 11) that we see becoming an ever more common and necessary element of our increasingly complex world, one in which sustaining innovation will remain a constant challenge.

Barcelona, Spain Steven MacGregor
Silicon Valley, USA Tamara Carleton

References

Carayannis E G, Campbell D F J (2009) "Mode 3" and "quadruple helix": Toward a 21st century fractal innovation ecosystem. *Int J of Technol Manage, 46*(3/4):201–234.

Carayannis E G, Campbell D F J (2010) Triple helix, quadruple helix and quintuple helix and how do knowledge, innovation and the environment relate to each other? A proposed framework for a trans-disciplinary analysis of sustainable development and social ecology. *Int J of Soc Ecol Sust Dev, 1*(1): 41–69.

Chesbrough H (2003) *Open innovation: The new imperative for creating and profiting from technology.* Harvard Business School, Cambridge, US.

Etzkowitz, H. (1996) The triple helix: Academic-industry-government relations in New York. In Raymond, S U (Ed.), *The technology link to economic development: Past lessons and future imperatives.* Academy of Sciences, New York.

Friedman T L (2005) *The world is flat: A brief history of the twenty-first century.* Farrar, Straus & Giroux, New York.

Ghemawat P (2011) *World 3.0: Global prosperity and how to achieve it.* Harvard Business Press, Cambridge.

Grayson D, Hodges A (2004) *Corporate social opportunity! Seven steps to make corporate social opportunity work for your business.* Greenleaf Publishing, Sheffield, UK.

Huston L, Sakkab N (2006) Connect and develop: Inside Procter & Gamble's new model for innovation. *Harvard Bus Rev*, March:58–66.

Norman D (1986) *The design of everyday things.* MIT Press, Cambridge, US.

Organisation for Economic Co-operation and Development (OECD) (1997) *National Innovation Systems.* OECD Publications, Paris.

Porter M E and Kramer M (2011) Creating Shared Value'. *Harvard Bus Rev*, January/February: 62–77.

Powell W W, Grodal S (2005) Networks of innovators. In Fagerberg J, Mowery D C, Nelson R R (Eds.), *The Oxford handbook of innovation.* Oxford University, New York.

Acknowledgments

We would first of all like to thank all the authors for their hard work in producing multiple drafts of extremely insightful chapters that allowed us to learn many new things in this exciting field. We also wish to thank Elias Carayannis, the editor-in-chief of this Springer series, as well as Charlotte Cusumano and Nick Philipson at Springer for their thoughtful guidance and patient support during the design, preparation, and production of this volume. On a personal note, our families have been equally understanding of the additional time and the heart an endeavor like this requires.

Contents

SECTION: VISIONS

SECTION II: RESEARCH

SECTION III: EXPERIENCES

Contributors

Laura Ahonen Administrative Centre, City of Jyväskylä, Finland

Álvaro Morón Alonso BBVA Group, Madrid, Spain

Andrea Bikfalvi Department of Business Administration and Product Design, University of Girona, Girona, Spain

Susanna Bill Department of Design Sciences, Lund University, Lund, Sweden

Javier Sebastián Cermeño Capgemini, Madrid, Spain

Amaresh Chakrabarti Innovation, Design Study and Sustainability Laboratory (Ideas Lab), Centre for Product Design and Manufacturing, Indian Institute of Science, Bangalore, India

Dagmar Chlosta Adidas Group, Schwaig, Germany

David Coates Growth Agenda Limited, London, United Kingdom

Roger Dennis Growth Agenda Limited, London

Tuija Hämäläinen Administrative Centre, City of Jyväskylä, Finland

Jenny Ingridsdotter Baltic and East European Graduate School, Södertörn University, Södertörn, Sweden

Tim Jones Growth Agenda Limited, London, United Kingdom

Andreas Larsson Department of Design Sciences, Lund University, Lund, Sweden

Céline Louche Vlerick Leuven Gent Management School, Leuven, Belgium

Pilar Marquès Department of Business Organization, Universitat de Girona, Girona, Spain

Nam Mokwunye, MBA, MFA Technology Transfer and Triple-Helix Innovation, Center for African Studies, Stanford University, Stanford, CA, USA

Annika Olsson Department of Design Sciences, Lund University, Lund, Sweden

Leo Roodhart Said Business School, University of Oxford, Oxford, United Kingdom

Nigel Roome Vlerick Leuven Gent Management School, Leuven, Belgium

Josep Lluis de la Rosa TECNIO – Centre, University of Girona, Girona, Spain

Alexandra Simon Department of Business Organization, Universitat de Girona, Girona, Spain

Gokula Vijaykumar A.V. Innovation, Design Study and Sustainability Laboratory (Ideas Lab), Centre for Product Design and Manufacturing, Indian Institute of Science, Bangalore, India

Chapter 1
Exploring a Vision for Sustaining Innovation in African Economies

Nam Mokwunye

Abstract There is an opportunity for African countries to capitalize on economic gains made in the telecommunication sector over the last decade and self-imposed pressure to achieve the United Nations Millennium Development Goals (MDGs). Thus far, African leaders have accepted capacity building and infrastructure development as two responsibilities which, when implemented, should accelerate internal and regional economic growth. But to accelerate development and overcome socioeconomic challenges associated with competing in an information-driven, flat economy, leadership in African governments, businesses and research (academic and nonacademic) must adopt a scalable and sustainable technology transfer agenda designed for multihelix innovation. Such an agenda would allow internal and external stakeholders to benefit from the network of productive information, transactions, and human capital that the continent's telecommunication operators have accumulated over the decade. With multihelix collaboration, these physical networks could become transformative platforms that make it possible to scale economies, increase security and crop yields, promote human rights and civil liberties, improve health service response, multiply transactions and Foreign Direct Investments (FDI), and accelerate intellectual content exports. The convergent telemedia platforms could fuse physical infrastructure with virtual networks, social applications, and human activity to sustainably galvanize innovation, generate opportunities, create value, and change lives. New business models such as pay-as-you-go and residual revenue sharing, which leverage the absence of legacy systems, can facilitate this fusion.

N. Mokwunye, MBA, MFA (✉)
Technology Transfer and Triple-Helix Innovation, Center for African Studies, Stanford University, 216 Encina Hall West, 94305-6045, Stanford, CA, USA
e-mail: diginam@gmail.com

S.P. MacGregor and T. Carleton (eds.), *Sustaining Innovation: Collaboration Models for a Complex World*, Innovation, Technology, and Knowledge Management,
DOI 10.1007/978-1-4614-2077-4_1,

Along with infrastructure and capacity building (Kagire 2011), technology transfer within a multihelix innovation framework is the third wheel of a nation-building tricycle upon which progressive collaboration between the state, market, and civil society relies. Thus far, leaders of African countries have accepted capacity building and infrastructure development as two responsibilities which, when implemented, should accelerate internal and regional economic growth.

Capacity building concerns the improvement of the systems, processes and resources required for the public workforce to adequately provide public services to a people and its private sector. The World Bank spent nearly $10 billion on capacity building between the years 1995 and 2004. Nine billion of the funds came in the form of aid. Now, imagine a nation with improving public sector capacity but without adequate infrastructure with which to execute it. According to an African Development Bank (AfDB) and World Bank study, the current state of infrastructure in African countries reduces economic growth by two percentage points and truncates business productivity by 40% (Africa Infrastructure Knowledge Program, AIKP 2009). In fact, African countries would collectively require $93 billion/year over 10 years to build enough federal and city infrastructure initiatives such as power, transportation (air, road, and rail), water, information and communication technology (ICT) continent-wide to achieve the Millennium Development Goals (MDGs) by 2015 (ibid.). This amount has escalated over the years as it was previously $45 billion/year, thus begging the question of the impact of timing on such infrastructure investments.

As material prices increase and existing infrastructure decays, so do overall project costs rise. As a result, some advocates, such as Abdoullah Wade, president of Senegal, Olusegun Obasanjo and Thabo Mbeki, former presidents of Nigeria and South Africa, respectively, have called for a continent-wide "Marshall-Plan" (Nugent 2001; Quist-Arcton 2001; Mucavele 2005). According to them, a "Total Action Development (TAD)" approach would identify all funding sources for the continent's total infrastructure needs in advance, then commence and complete projects and payments within 5 years, as did the Marshall Plan for post-WWII Europe between 1947 and 1951.

However, most governments in African countries have, in the recent past, funded capital-intensive projects as public–private partnerships (PPP) (Cassel et al. 2010), a system similar to that which has succeeded since 1993 in Canada where it is called "P3" (Canadian Council for Public–Private Partnerships 2011). Still, the question of whether "aid" (Sachs and McArthur 2009) or "trade" is best for infrastructure development and capacity building remains. The intellectuals and activists addressing this issue can be grouped into three camps: (1) those who support aid primarily such as Bono (Nugent 2001; Alagna 2008) or Jeffrey Sachs (Sachs 2005), (2) those who support trade primarily such as William Easterly (Postrel 2006; Easterly 2006) or Moyo (2009), and (3) those who advocate trade backed by aid managed with vision and discipline such as Ngozi Okonjo-Iweala (Okonjo-Iweala 2007).

While Moyo makes a strong case that $1 trillion (Moyo 2009) in developmental aid was poorly placed and badly managed, Okonjo-Iweala sees both as necessary and points to her restructure and pay-down of Nigeria's debt from $33 to $6 billion within 15 months between 2004 and 2005 while she was Nigeria's first female finance minister (Center for Global Development 2005). Paying the debt reduced interest expenses and allowed the country to attract cheaper debt and equity financing necessary to execute capital-intensive infrastructure and capacity development initiatives. For Africa's most populous (151 million) country, largest oil producer, and second largest economy, the result has been a recently oversubscribed $500 million Eurobond whose yield encouraged Angola to issue the same.

Regardless of methodology and timing of financing, change is occurring in many African countries and, in the words of Kofi Annan, "What was termed 'the hopeless continent' ten years ago has now unquestionably become the continent of hope" (Africa Progress Panel 2011). For this reason, stakeholders are optimistic about the near and distant future of the African continent. Now is the time to develop a comprehensive technology transfer agenda.

1 What is Technology Transfer?

Technology transfer (TT) in its basic form is the transfer of concepts, ideas, knowledge, processes, trade secrets, inventions, and devices—collectively called "technologies"—from one body, entity, organization, or nation to another. The literature discussing the many dimensions of the TT field is significant. TT includes the technology item, the developer, the transfer channels, and the recipient. Saggi (2004) introduces policies and multilateral concerns and then proposes ways to encourage international technology transfer (ITT) to developing countries. Parente and Prescott (1994) discuss the correlation between trade, foreign direct investment (FDI), and TT costs, while Garduno (2004) considers TT metrics for developing countries. Johnson et al. (1997) discuss the transfer of content from the inventing environment into the adoption environment and then diffusion within the adopting environment. Rogers and Mills (1989), as well as Johnson and colleagues, posit that whereas TT refers to the development and transfer of technologies from one place to another, diffusion is the spreading or accepted use of those technologies within a society, organization or amongst a people (Rogers 1995). Thus, technologies are transferred while innovations are diffused. This dynamic, multilateral development–transfer–diffusion process, ideally, should self-perpetuate and renew its life cycle many times over decades.

A suboptimal transfer fails to renew its life cycle, is demand-driven, and as a result, does not experience its full commercial potential. One of the reasons for this is the absence of collaboration between government, industry, and researchers. This happened in the telecommunication industry of many African countries in which policy, license auctions, and transactions fees engaged government with industry, thus leaving universities and research out of the equation. In contrast, an optimal transfer would capitalize on the policy, commercial, and research capabilities of the government, industry and academic and nonacademic research institutions, respectively, to develop new product and business life cycles. These new life cycles establish new supply–demand equilibriums, often for different markets, thus dispersing the innovations at a more rapid pace to contribute to regional socioeconomic development.

A unidirectional transfer is made from one entity to another without reciprocation, and a bidirectional or multidirectional transfer is made, reciprocally, between two or more entities. While there is very little literature about the latter type of transfer, a creative mind might imagine that optimal transfers are more likely when there is mutual benefit in a transaction. For example, the funds, technology, expertise, and human capacity transferred to Kenyan telecom provider Safaricom from a consortium that included the UK Department for International Development (DFID), Sargentia, Vodafone, IBM and the Kenyan government, produced for the world M-Pesa, a leading mobile transaction platform that introduced efficiencies, inventions, transfers, and diffusions in both the banking and telecommunication industries through "branchless" banking on the African continent and worldwide (Hughes and Lonie 2009).

2 How Many Players in the Innovation and Technology Transfer Helix?

In discussing triple-helix innovation, Etzkowitz and Leydesdorff (1995, 2000) first posited that where the relationship between government, industry, and *entrepreneurial universities* (Etzkowitz 2008) was once linear and predictable, it is now a helix in which the three players retain genetic components of the others. TT relates to innovation in that the transfer of ideas, knowledge, inventions, processes, and technologies, as TT is defined, is the bedrock of innovation which is the backbone for diffusion and impact within a society. In a digital economy, the quadruple-helix more accurately proposes that civil society (developmental organization, nongovernmental organizations, and civil groups) now play a critical role in the localized innovation, transfer, and diffusion of technologies for economic development (Carayannis and Campbell 2009). In this age of globalization in which government and industry are converging, research is becoming less formal and localization gains value (Ghemawat 2011) to service providers and end-users, the quadruple-helix model as depicted by Carayannis and Campbell (2009) is a more appropriate representation of the interaction of the agents of innovation than is triple-helix.

But just as the inclusion of civil society is not obvious in the triple-helix model, that of the end-user is not obvious in the quadruple-helix model. It appears, instead, that in making their argument, authors such as Arnkil et al. (2010) erroneously amalgamate civil society and end-users, which, while growing together in influence as the drivers of the helix relationship, are in fact not the same. As telemedia simplifies thought-sharing and collective engagement, the end-users are becoming the researchers, exercising more power over the physical and conceptual product, and thus gaining more influence over government and industry. The recent user-driven civil uprisings in North Africa and the Middle East stand as examples. Civil society aids and facilitates this change process, advocating on the end-user's behalf. This is embodied in the role developmental organizations such as the World Intellectual Property Organization (WIPO), World Trade Organization (WTO), International Telecommunication Union (ITU), World Health Organization (WHO), the International Monetary Fund (IMF), The World Bank, and The Bill & Melinda Gates Foundation continue to play in the successful transfer of technologies in emerging markets. This is particularly the case in Africa in which these organizations, in some cases, write the framework for government policies that impact populations.

So, it is possible that the innovation and technology transfer helix may involve five distinct players, a quintuple-helix, as recently proposed by Carayannis and Campbell (2010), or even more, as Leydesdorff (2010) proposes with the algorithm *N-tuple*-helix, a composition in which he acknowledges Carayannis and Campbell. The graphics would be more representative of the ongoing dynamism of global geopolitical and geoeconomic interaction.

TT discussed within a multihelix framework refers to transfers of intellectual assets that stimulate national or regional economic development within the helix that includes end-users, government, industry, researchers (formal or informal), and civil society (domestic or international). Such a mix of policy, research, and production has been used to resuscitate ailing economies throughout recent history, for example in post-World War II Japan and Germany, in the Asian Tigers (1960s–1990s) that included South Korea, and more recently in the Southeast Asian Tigers (Afrol.com 2010). This provides evidence of the efficacy of TT applied within a multihelix framework.

Though some African countries (such as Mauritius and Botswana) are moving in the right direction, others still need to put much in place including a value for research, alignment of policy with economic and sociocultural objectives, and the inclusion of the end-user in the vision-creation and decision-making. Such a bold agenda cannot be achieved without the buy-in and commitment of leadership.

3 Maximizing the Multihelix Technology Transfer Agenda

To accelerate economic growth, leadership in African governments, businesses, and research organizations (formal and otherwise) must adopt a scalable technology transfer agenda designed for multihelix innovation because, while government and industry will continue to play the parts they have in growing sectors, the participation of researchers and civil society will be necessary to find new business life cycles and accelerate economic growth. African countries have progressed significantly over the last 20 years by committedly developing their information and transaction systems. The result has been improvements on key development indicators in the economic, health, agriculture, and education sectors. According to the Africa

Progress Report (Africa Progress Panel 2011), "In the five years before the 2008 financial crisis, Africa grew faster than most other world regions, with more than 40 percent of its countries enjoying an average annual GDP growth rate of five percent or more." This should make African countries an attractive investment target for fair-minded, yield-seeking investors. And in some cases it has.

From 1990 to 2009, per capita gross national income (GNI) increased from $566 to $1125; life expectancy at birth increased from 50 to 53; primary education completion rate went up from 51% (1999) to 64% (2008); and while the world's per capita CO^2 emission went up from 4.3 to 4.6 mt (2007), the African continent went down from 0.9 to 0.8 (2007) based on World Bank (2010) data. Despite the progress, the issues impacting the common woman, man, and child in African countries still persist at alarming rates. For example, over 765 million persons are still at risk for malaria (The World Health Organization 2010), and 22.5 million (2.25%) of the African population still lives with HIV/AIDS (UNAIDS.org 2010; The World Health Organization 2010).

Technology transfer matters because it has contributed to the aforementioned progress and can also contribute to overcoming persistent challenges. As an asset, technology enables individuals, organizations, industries, and governments to generate value from more basic assets. Thus, it stands to reason that the proper application of technologies—when transferred to locations and situations of need with adequate support—will help scale economies, increase security and crop yields, promote human rights and civil liberties, improve health service response, multiply transactions (Africa Progress Panel 2011) and Foreign Direct Investments (FDI), and accelerate intellectual content exports. Yet, until recently, technology transfer has been given cursory consideration in the redevelopment of African countries, people, social structures, polity, and, ultimately, economies.

To maximize collaborative efforts between state, markets, and civil society, which could accelerate development and overcome socioeconomic challenges associated with competing in an information-driven, flat (Friedman 2005), and local (Ghemawat 2011) economy, leaders in African governments, industry, and universities must adopt a scalable TT agenda designed for multihelix innovation. Unlike with past transfers that were dominated by multinationals and intergovernment exchanges, however, the transfer of tomorrow would permeate beyond the scenario plans of the largest corporations to be available to even the smallest vendor and most isolated individual. The collaboration between policy, research and industry could disrupt existing systems and create a paradigm shift that will produce new life cycles—productive layers—for activities associated with government, university and industry as well as with civil society and individual citizens.

The new layers of physical and virtual networks are social instruments that now permeate and support general and specific growth in African countries, making such a socioeconomic transformation possible. And the participation of academia and civil society, through research (academic and otherwise) and advocacy, will be necessary to allow industries to discover, develop, and sustain new business life cycles which they can scale to accelerate economic growth.

4 Telecom Networks as Physical Foundation for Continental Virtual Networks

In the last decade, the telecommunication (telecom) industry has grown faster on the African continent than in any other part of the world while developing physical and social networks using varying levels of public–private alliances. Nearly $20 billion in foreign direct investment (FDI) pushed the telecom industry from 20–174 to 333 million subscribers at the end of 2000, 2007, and 2010, respectively (International Telecommunication Union 2010). Given that telecom made up 10% of the $235 billion in GDP growth across the continent between 2001 and 2007 (McKinsey Quarterly 2009), the African telecom market was the fastest growing telecommunication market in the world. During that growth period, the technology transfer occurred in five steps: registration, establishing regulatory infrastructure, privatization, adoption, and diffusion.

4.1 Registration (1910-Recent)

This step of the transfer started as far back as 1910 when South Africa became a listed member state of the International Telecommunication Union (ITU), the governing body for global issue in telecom located in the United Nations. All but one country, Libya, are listed as a member state of the ITU.

4.2 Establishing Regulatory Infrastructure (1996–Recent)

Most governments demonstrated political will by creating a regulatory framework that conformed to the standards set by the ITU for privatization. This step involved the following: (1) separating the regulatory, privatization, and policy functions by creating independent oversight bodies and (2) building new capacity to handle projected demand explosions by engaging ITU experts to train employees or partaking in exchange programs with more experienced counterparts in more experienced ITU-member countries.

4.3 Privatization (1998–Recent)

This step involved the government auctioning the right and license for private operators to utilize radio frequency spectrum within their national borders to build physical networks and sell cellular voice services to business and consumer end-users. In most countries, it was a PPP of sorts for which the government would continue receiving royalties. Pressured by the demand of private sector money and the desire to exhibit a trickle-down progress for discouraged populations, privatization became a panacea, a default method for generating competition and driving development with the telecom sector (Klein and Gray 1997). As a result, new companies and brands of global industries were created. For example, MTN was a six million subscriber operator in South Africa before it purchased a GSM license in Nigeria for $285 million in 2001. After 10 years of operating in Nigeria, MTN's global subscriber base is now 142 million, 42 million of whom are in Nigeria. It has since declined a $23 billion acquisition attempt from India's Bharti telecom and remains the largest telecommunications company in Africa.

Despite such success stories, experts such as Kikeri and Phipps (2007) question the significance early (auction-based) privatization plays in economic improvement and suggest that other forms of deregulation strategies might be equally effective within an enabling policy and implementation environment (Naceur et al. 2008). Diverse findings imply that the GSM boom may have been, on aggregate, a suboptimal technology transfer that enriched a few license-holders across the continent, while the government did not maximize the full value of the spectrum licenses on behalf of their people in what Sachs et al (2000) call a mere "change of ownership" (Sachs et al. 2000). Klein and Gray (1997) conclude that there are many ways to privatize a complex network based on the type of network, timing, and commitment of leadership.

Such learning points combined with a desire to cultivate their own network and intellectual infrastructure have driven Ethiopia to express no intention to privatize in the immediate future, despite pressure from World Bank and telecommunication expansionists. The East African country has the continent's third largest population (80 million), a low corruption index reading, and one of the continent's most committed and consistent leaderships who has tied all government targets to achieving the MDGs by 2015. In contrast to Nigeria, South Africa, Egypt and Kenya, Ethiopia has instead adopted a privatization strategy, which keeps the frequency spectrum in the hands of the government as the protector of the end-user's interests, builds and owns the network, and issues licenses to resell and develop new telecom services that will run over the Ethio Telecom network. Over a 6-year period, the government intentionally rolled out media-capable voice infrastructure to cover 90% of the rural population, broadband infrastructure to meet municipal needs in nine regions of the country, and e-government infrastructure to meet the needs of its major research centers.

To introduce the fiduciary discipline of the private sector, the government executed a management contract with France Telecom whose pay would hinge upon its ability to meet preagreed targets over a 2-year period. Ethio Telecom, which had eight million subscribers in December 2010, currently has 11 million with a 32 million target for 2015. Should Ethiopia meet this target they could auction rights of access to their network, and generate as much as $11 billion, more than Nigeria has generated from early auctions since 1998. Alternatively, Ethio Telecom could also generate significant amounts through revenue sharing with applications developers who could also charge a nominal access fee. Such a windfall for Ethiopia would be greater than their 2009 national budget and adequate to pursue projects that move it closer to achieving the MDGs.

4.4 Adoption (1998–Recent)

Previous attempts had been made by such companies as Thuraya and Iridium to introduce satellite mobile devices to the African continent in the 1990s. However, prices were too steep, and equipment was too bulky to address the needs of the masses. This was not the case with GSM. By the time the newly built GSM 900 and 1,800 MHz networks turned on in 2000, basic phones sold at a price below $200 and SIM cards, and thus subscriptions, sold for $150 in Nigeria. By 2010, when ubiquitous penetration was evidenced at over 300 million [30% of Africans[1] (Vital Wave Consulting 2011)], SIM cards sold for no more than $3 and in some countries was part of bundled services. According to Mobile Monday, there were 500 million subscribers in 2011 (Mobile Monday 2011).

Indigenous operators succeeded in building physical networks across the African continent. While the Nigerian, Egyptian, Ethiopian, Kenyan and South African air flight providers made it easier for Africans with means to move from country to country, the telecom network providers reduced their need to travel at all and took it a step further: the physical networks brought common people together with a force similar to the advent of the modern train in nineteenth century America or the radio in pre-World War II Great Britain. Africa's virtual networks were born and the way was paved to activate social networks.

4.5 Diffusion (1998–Recent)

With penetration came diffusion of methodologies—how people used the technologies—and innovations that reverberated both on and off the continent. Operators such as Globacom (Nigeria), Celtel (Sudan), Safaricom (Kenya), Econet (Zimbabwe), MTN (South Africa), and Vodacom (South Africa) expanded to become Pan-African and multinational. Original investors such as Actis (Celtel) and the Kenyan government (Safaricom) exited in some of the biggest M&A and IPO transactions in emerging market history. But much of the growth of these physical networks was demand-driven, and thus suboptimal. The people needed the service and were willing to pay. Product, management, agent, finance, and timing were aligned. Thus, it is possible that, in an enabling environment, anyone with money, experience, and a license may have had an equally successful brand.

---

[1] Vital Wave Consulting and the Bill & Melinda Gates Foundation learned from MoTech that only one third of their village mobile device users needed to have phones for the device to be used by most, if not all, participant in their program in which participants used mobile devices to update data about pregnant mothers.

5 Physical and Virtual Networks as Foundation for African Social Network

So, when Elana Berkowitz, an innovation advisor in the US State Department, said, "Africans are behind some of the most effective digital tools for driving social change and economic inclusion…." (Mobile Monday 2011), she was referring to a new optimal growth era, driven by informal and formal research, in which physical network expansion would catalyze diffusion, enable virtual networks, and galvanize social networks within and between African countries and the rest of the world. A technology transfer agenda, at this point, would allow internal and external stakeholders to benefit from the network of productive information, transaction, and human capital that the continent's telecommunication operators and consumers have accumulated over the decade.

The physical and virtual networks across the African continent have now become the foundation for an African social network—networks supporting and enabling the daily virtual activities of business people, legislators, researchers (academic and nonacademic), civil society, and everyday consumers within African countries and across the African continent. These networks could become social platforms that transform the African human network—a valuable sliver of Africa's network of productive information, transaction, and human capital, anchored upon 500 million subscribers—into an extensive social network that penetrates villages and reaches global regions from where it returns ideas, people, and money. Such a technology transfer agenda would make it possible to scale economies, increase security and crop yields, promote human rights and civil liberties, improve health service response, multiply transactions and Foreign Direct Investments (FDI), and accelerate intellectual content exports. This would be the new development paradigm, an engine made in Africa, by Africans, for Africans to further benefit the world.

Also, with coordinated regulation and legislation such as those concerning roaming, number porting, SIM card registration, copyright, patents and licensing, the rights of the end-user would finally become paramount. As such, these developments would also dovetail ideally with new social networking platforms, such as Facebook, life-saving platforms such as Ushahidi and MoTeCH, and transaction-enabling platforms such as Paga and MXit. When these social innovations combine with the depth and breadth of the physical, virtual, and human networks, they brazenly expose the creativity of the African technologist, entrepreneur, and digital service consumer—without bias.

6 New Business Models Can Change Stakeholder Interactions and Benefit End-Users

The proliferation of physical wireless networks in African countries exposes opportunities to introduce business models that change the way business is done and humans interact—globally. Networks are both physical and social and can accumulate value when operated upon platforms that empower merchants and consumers. A progressive technology transfer agenda can encourage these new business models that can produce surprising results that may impact economic growth. New internet standards, such as HTML5 (McKinsey Quarterly 2011) and IPv6, will allow African countries to continue bringing leverage to their vast physical and social infrastructure, unencumbered by legacy systems, to drive innovation.

Whereas the first, second, and third generation mobile transfers were unidirectional with demand-driven and suboptimal growth, the current and future generation transfers should be bi- and multidirectional and be driven by research. This should especially be true as African opportunities attract investors, operators, and adopters of innovative applications with diverse uses. The interactions would result in a cascading evolution of new business models that should eventually benefit all stakeholders by facilitating a fusion between service providers and consumers, increasing penetration and customization and reducing costs and pricing. The following are some examples.

Pay as you go (PAYG), or prepay, still accounts for over 80% of mobile service revenues on the African continent (Nwosu 2011). The figures are reversed in the USA with billed services accounting for most telecom revenues. PAYG made scratch cards a viable business for not only telecommunication credit agents and distributors but also branchless bankers (Kubzansky et al. 2011). All these concepts, already entrenched in African countries, are now being applied globally.

The next generation 4G convergent mobile telemedia platforms, such as Galaxy Wireless (Nigeria), XS Broadband (Nigeria), and Afrimax (The Netherlands) which has procured and are implementing 4G licenses in African countries to cover over 300 million people, are considering developing a transcontinental collaborative broadband network with low-cost pay-per-use business models that fuse physical infrastructure with virtual networks, social applications and digital human activity to sustainably galvanize innovation, generate opportunities, create value, and change lives.

Along the same line, the Kenyan telecom regulators have already announced a reduction in their fee for 3G spectrum to $10 million, thus comparing favorably to $80 million in Tunisia, $150 million in Nigeria, and $2.5 billion in India, and $7 billion in the UK. The hope for such low-barrier-to-entry licensing strategies is that operators may provide superior quality services at affordable prices while incubating partnerships with application developers to accelerate adoption, innovation, and diffusion.

Further, social video platforms, such as the start-up PublicVine, will allow African producers of video content to export to consumers around the world using their mobile phone and their extensive social network. PublicVine's pay-per-view subscription and residual content revenue sharing model is intended to monetize social media, increase financial return to content owners and allow subscribers targeted content choice and increased privacy.

Finally, the success of transfers, diffusions and local innovations will be proven by successful investments and transformation of lives. Private equity firm Actis, benefitted significantly from the trade sale of Celtel. Recently, Tim Draper, Founder and Managing Director of Silicon Valley venture capital firm Draper Fisher Jurvetson, has invested in Paga, the Nigerian mobile banking venture. These investments follow the early risks taken by Mark Shuttleworth, the South African inventor of the Ubuntu open source operation system, and Ayisi Makathiani, the Kenyan founder of Africa Online. They are all proof that money cyclically follows innovation produced by technology transfer. In the end, African peoples could be the beneficiaries.

7 Conclusion

Innovations such as the aforementioned will be instrumental for enabling Africa's telecom industry to deliver superior service to end-users. While frequency spectrum will become like "gold," a data-centric agenda will be fulfilled with 4G mobile technologies such as WiMAX and LTE connected by undersea and terrestrial fiber networks such as GloOne, MainOne, SEACOM, and SAT3 or by celestial networks such as O3b.

While the near-term business case will be made by convergence of SMS, voice, video, and social networking on IP social telemedia platforms utilizing business models that make physical networks scalable and sustainable, the long-term socio-economic case will be made by the quality of collaborations between end-users, government, industry, researchers (academic and nonacademic), and civil society. This will produce new life cycles that enable the internal innovations and external transfer of tomorrow's technologies and produce economic growth to the benefit of all peoples on the African continent.

References

Africa Infrastructure Knowledge Program (AIKP) (2009) AICD Reports: Africa's Infrastructure: A Time for Transformation. http://www.infrastructureafrica.org/aicd/library/doc/552/africa%E2%80%99s-infrastructure-time-transformation. African Development Bank Group, Tunisia

Africa Progress Panel (2011) Africa Progress Report: The transformative power of partnerships. http://www.africaprogresspanel.org/files/7713/0441/3939/APP_APR2011_FINAL.pdf. African Progress Panel, Geneva

Afrol.com (2010) Can Africa really learn from Korea? http://www.afrol.com/articles/22953

Alagna J (2008) Bono's DATA, Rocking the Development World. http://www.devex.com/en/articles/bonos-data-rocking-the-development-world. Devex, Washington, DC

Arnkil R et al. (2010) Exploring Quadruple HelixOutlining user-oriented innovation models: Final Report on Quadruple Helix Research for the CLIQ project. Tampereen yliopistopaino Oy Juvenes Print, Tampere 2010

Canadian Council for Public–Private Partnerships (2011)—CCPPP. http://www.pppcouncil.ca/

Carayannis E G and Campbell F J (2009) "Mode 3" and "Quadruple Helix": Toward a 21st Century Fractal Innovation Ecosystem. International Journal of Technology Management 46 (3/4), 201–234

Carayannis E G and Campbell F J (2010) Triple Helix, Quadruple Helix and Quintuple Helix and How Do Knowledge, Innovation and the Environment Relate To Each Other? A Proposed Framework for a Trans-disciplinary Analysis of Sustainable Development and Social Ecology. International Journal of Social Ecology and Sustainable Development (IJSESD), 1(1):41–69

Cassel C et al. (2010) Building African Infrastructure with Chinese Money. http://www.barcelonagse.eu/tmp/pdf/ITFD10Africa.pdf
Center for Global Development (2005) Resolving Nigeria's Debt Through a Discounted Buyback. http://www.cgdev.org/section/initiatives/_archive/nigeriandebtrelief
Easterly W (2006) The White Man's Burden: Why the West's efforts to aid the rest has done so much ill and so little good. Penguin Press, New York
Etzkowitz H (2008) The Triple Helix: University-Industry-Government Innovation in Action. Routledge, London
Etzkowitz H and Leydesdorff L (1995) The Triple Helix—University-IndustryGovernment Relations: A Laboratory for Knowledge-Based Economic Development. EASST Review 14, 14–19
Etzkowitz H and Leydesdorff L (2000) The dynamics of innovation: from National Systems and "Mode 2" to a Triple Helix of university–industry–government relations. Reseach Policy, 29:109–123
Friedman T L (2005) The World Is Flat: A brief history of the twenty-first century. Farrar, Straus & Giroux
Garduno, E (2004) Rethinking TT Metrics for Developing Countries. http://iipi.org/wp-content/uploads/2010/07/TTMetrics.pdf. International Intellectual Property Institute, Washington, D.C.
Ghemawat P (2011) World 3.0: Global Prosperity and How to Achieve It. Harvard Business Press, Cambridge
Hughes N and Lonie S (2009) M-Pesa: Mobile money for the unbanked. http://www.gsmworld.com/documents/INNOVATIONS-GSMA_FINAL-01-22-09.pdf, Innovations, pp. 76–96
International Telecommunication Union (2010) Key Global Telecom Indicators for the World Telecommunication Service Sector. http://www.itu.int/ITU-D/ict/statistics/at_glance/KeyTelecom.html. ITU, Geneva
Johnson S et al. (1997) Expanding the Content Base of Technology Education: Technology Transfer as a Topic of Study. Journal of Technology Education, pp. 35–49
Kagire E (2011) Capacity building in Africa transcends MDGs. http://sweden.embassy.gov.rw/content/view/203/98/. Rwandan Embassy in Sweden
Klein M and Gray P (1997) Note No. 104: Competition in Network Industries—Where and How to Introduce It. The World Bank and International Finance Corporation, Public Policy for the Private Sector, http://rru.worldbank.org/documents/publicpolicyjournal/104klein.pdf
Kikeri S and Phipps V (2007) According to Note No. 321: A Record Year for Initial Public Offerings in 2007 The World Bank and International Finance Corporation, Public Policy for the Private Sector, http://rru.worldbank.org/documents/publicpolicyjournal/321Kikeri_Phipps.pdf
Kubzansky M et al (2011) Promise and Progress: Market-based solutions to poverty in Africa. Monitor Group, Johannesburg
Leydesdorff L (2010) The Triple Helix, Quadruple Helix, ..., and an N-tuple of Helices: Explanatory Models for Analyzing the Knowledge-based Economy? http://www.leydesdorff.net/ntuple/ntuple.pdf
McKinsey Quarterly (2009) What's driving Africa's growth http://www.mckinseyquarterly.com/Whats_driving_Africas_growth_2601
McKinsey Quarterly (2011) How new Internet standards will finally deliver a mobile revolution http://www.mckinseyquarterly.com/Telecommunications/How_new_Internet_standards_will_finally_deliver_a_mobile_revolution_2788
Mobile Monday (2011) Mobile Africa Report: Regional hubs of excellence and innovation. http://www.mobilemonday.net/reports/MobileAfrica_2011.pdf, Helsinki
Moyo D (2009) Dead Aid: Why aid is not working and how there is a better way for Africa. Farrar, Straus, Giroux. New York
Mucavele F G (2005) The New Partnership for Africa's Development (NEPAD). http://siteresources.worldbank.org/INTAFRREGINICOO/Resources/NEPAD_Brief.pdf. The World Bank, Washington, DC

Naceur S B et al. (2008) Does stock market liberalization spur financial and economic development in the MENA region? http://www.sciencedirect.com/science?_ob=ArticleURL&_udi=B6WHV-4RDPYPW-2&_user=10&_coverDate=12%2F31%2F2008&_alid=1751975550&_rdoc=2&_fmt=high&_orig=search&_origin=search&_zone=rslt_list_item&_cdi=6860&_sort=r&_st=13&_docanchor=&view=c&_ct=4&_acct=C000050221&_version=1&_urlVersion=0&_userid=10&md5=524e25693e919f64aba4412c7b7b606c&searchtype=a. Journal of Comparative Economics, Volume 36, Issue 4, December 2008, Pages 673–693, special symposium on East Germany

Nugent B (2001) Can Bono Save The World? http://www.time.com/time/covers/1101020304/story.html. Time, New York

Nwosu G (2011) Discussions with CEO and CTO of Galaxy Wireless. Abuja

Okonjo-Iweala N (2007) Closing TED Conference. http://www.ted.com/talks/ngozi_okonjo_iweala_on_aid_versus_trade.html. TEDGlobal, New York

Parente S and Prescott E (1994) Barriers to Technology Adoption and Development. The Journal of Political Economy. 102(2):298–321. http://faculty.wcas.northwestern.edu/~lchrist/papers/barriers.pdf

Postrel V (2006) 'The White Man's Burden,' by William Easterly: The Poverty Puzzle. http://www.nytimes.com/2006/03/19/books/review/19postrel.html. The New York Times, Books Section

Quist-Arcton O (2001) A "Continental Strategy" for Building Africa Infrastructure Needed—Wade. http://allafrica.com/stories/200102080307.html?page=2. AllAfrica.com, Washington, DC

Rogers E and Mills T (1989) Co-citation Analysis of the Scientific Literature of Innovation Research Traditions: Diffusion of Innovations and Technology Transfer. Science Communication December 1989 11:181–208

Rogers E (1995) Diffusion of Innovations. Free Press, New York

Sachs et al. (2000) "The gains from privatization in transition economies: Is 'change of ownership' enough? Consulting assistance on economic reform II discussion paper". Vol. 63. Center for International Development, Harvard University, Cambridge, Massachusetts

Sachs J (2005) Why Aid Does Work. http://www.unmillenniumproject.org/documents/BBCJDSWhyAidWorks.pdf. BBC News, London

Sachs J and McArthur J W (2009) Moyo's Confused Attack on Aid for Africa. http://www.huffingtonpost.com/jeffrey-sachs/moyos-confused-attack-on_b_208222.html. Huffington Post, Washington, DC

Saggi K (2004) International technology transfer to developing countries. Economic Paper 62. Commonwealth Secretariat

UNAIDS.org (2010) Global Report: UNAIDS report on the AIDS epidemic. http://www.unaids.org/globalreport/documents/20101123_GlobalReport_full_en.pdf. Joint United Nations Programme on HIV/AIDS (UNAIDS)

Vital Wave Consulting (2011) mHealth in Ethiopia: Strategies for a New Framework. Bill & Melinda Gates Foundation, Seattle

The World Bank (2010) Data. http://data.worldbank.org/region/SSA. The World Bank, Washington, D.C.

The World Health Organization (2010) World Malaria Report 2010. http://www.who.int/malaria/world_malaria_report_2010/worldmalariareport2010.pdf. The World Health Organization, Geneva

The World Health Organization Media Center (2010) Millennium Development Goals: progress towards the health-related Millennium Development Goals. http://www.who.int/mediacentre/factsheets/fs290/en/index.html. The World Health Organization, Geneva

Chapter 2
CLIQ: A Practical Approach to the Quadruple Helix and More Open Innovation

Laura Ahonen and Tuija Hämäläinen

Abstract This chapter contains the lessons learned from a European innovation project called Creating Local Innovation in a Quadruple Helix (CLIQ). It starts by introducing CLIQ and its aims and then discusses the search for a Quadruple Helix innovation ecosystem by 16 partners, many of them medium-sized towns of Europe. It outlines the roles and challenges of the local authorities in innovation, describes various ways of collective learning, and introduces the end products of CLIQ, all planned to tackle issues such as measuring innovation, transfer of good practices, and ideal conditions for innovation to flourish. At the time of writing, the project is still in progress, yet the reader should get an idea of the main body of learning: what was learned from the research commissioned by CLIQ and through the comprehensive, pragmatically oriented work plan of the project. The learning, or legacy vision of CLIQ, is summarized at the end of the chapter.

> Some of those present can still clearly recollect the moment several years ago, when someone in a very tentative and uncertain manner for the first time said: quadruple helix? All the others in the room, in deep contemplation, tasted the words in their mouths, and slowly the thought cleared in their minds...you could almost see it materialize. The expression on peoples' faces brightened; they looked each other in the eye; some of them smiled a bit: yes, that's it! After being born, the thought never left us. "Quadruple Helix" – what on earth may it mean? We certainly did not know. Neither did we know how far this thought would take us to date.

1 What is CLIQ?

CLIQ is an acronym for "Creating Local Innovation in a Quadruple Helix." With effort and quite a lot of good luck, it became a European project of 16 partners under the Interreg IVC funding program of the European Union. The project partners are

L. Ahonen (✉) • T. Hämäläinen
Administrative Centre, City of Jyväskylä,, PL 193, Fin-40101 Jyväskylä, Finland
e-mail: Laura.Ahonen@jkl.fi

S.P. MacGregor and T. Carleton (eds.), *Sustaining Innovation: Collaboration Models for a Complex World*, Innovation, Technology, and Knowledge Management,
DOI 10.1007/978-1-4614-2077-4_2,

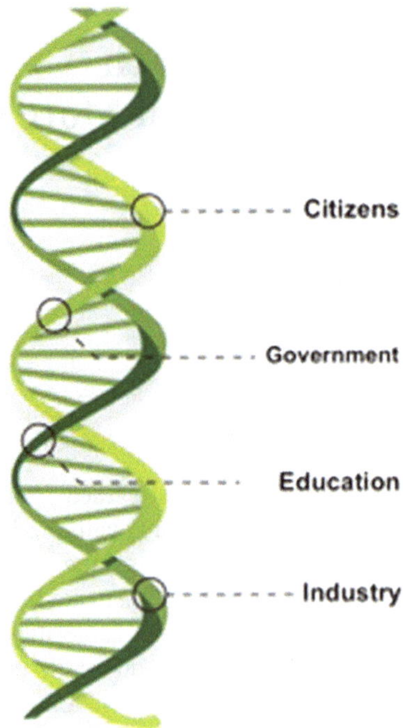

Fig. 2.1 Quadruple helix

local and regional authorities (many of them medium-sized towns), business and innovation centers, and chambers of commerce from ten European countries, which represent different roles and positions within the innovation chain. The Interreg program is about the exchange of knowledge and experience across Europe and is implemented under the European Community's territorial cooperation objective and financed through the European Regional Development Fund. The areas of support are innovation and the knowledge economy, environment and risk prevention (Interreg IVC Innovation and Environment. Regions of Europe Sharing Solutions 2010).

CLIQ had an ambitious work plan implemented for 3 years in 2008–2011. The long-term aim of the project is to optimize the benefits of globalization and innovation to small and medium sized enterprises (SMEs) and entrepreneurs in medium-sized towns. The main objective of the project is to strengthen local and regional authority policy and their capacity to support innovation more effectively.

The metaphor used at the beginning of the project to illustrate Quadruple Helix was derived from genetics: a DNA chain, where the different "helices" vine around each other and work together to the same purpose yet still maintaining identity (see Fig. 2.1). A paradigm shift toward user-driven and open innovation systems is seen to change the rules of play and the roles of the players, specifically: research, industry, government, and citizens. This chapter collects the learning from CLIQ and represents a pragmatic approach to the Quadruple Helix model of innovation: what could it mean in practice, and could it possibly bring anything new to the present discussion of open and sustainable innovation?

1.1 CLIQ and European Innovation Discourse

The Europe 2020 strategy names innovation as one of the keys to smart, sustainable and inclusive growth that is aimed at the whole continent (Communication of the European Commission 2010). The discussion of innovation systems, however, has not brought many radically new points of view to the scene for a long time. The notion of Triple Helix was innovative when introduced, but many practical examples show that there are inherent difficulties in Triple Helix. The world has changed since then. Nowadays, the dedicated knowledge institutions (universities, R&D institutions, think tanks, etc.) are not sole producers of knowledge. Many big international companies maintain large R&D departments, own a great number of patents, and are often able to hire the best brains of a generation with higher financial incentives than public universities. The public sector as a whole is suffering from continuous financial cuts and decreasing human resources. It cannot absorb innovations produced for many reasons, one of them being the very stiff system of public procurement, originally meant to make the use of the taxpayers' money more transparent and its users better accountable—a good aim that brought about an unwanted breed, if we may say. When talking about service innovation, obstacles include the lack of knowledge of customers' needs and the absence of a proper mindset for renewal.

In addition, the whole discussion of open innovation—so popular during the last few years—has brought into daylight the fact that the Triple Helix is incapable of incorporating large amounts of equal "innovators" in the system. Something more or different is needed. There are various attempts of doing so: see for example the classification in the Quadruple Helix research (Arnkil et al. 2010) commissioned by CLIQ, which sets out from the assumption that *civil society* is the missing element that needs to be engaged in innovation. Others have argued that the *financiers* are the fourth pillar of the Quadruple Helix. Which one of the various options presented is most effective to grasp the nature of more inclusive or (even) open innovation remains to be seen.

1.2 Innovation and Public Authorities

It goes without saying that the public sector is in need of more innovations. Facing a rapidly aging Europe and ever more scarce resources, the public sector needs to adopt new and more efficient ways to work. The new nature of innovation is reshaping public policy. In the innovation economy, a more responsive public sector and a comprehensive set of policies are in the center when gaining benefits from the changed situation. However, the challenges that the public sector is facing can work also as an innovation driver (OECD Committee for Industry and Innovation 2009). The public sector can have a wide variety of means to influence innovation demand also locally, even though national differences exist in view of the role and capacity of local authorities.

The public sector can directly influence its own *demand*. Public procurement could, in principle, increase innovation both in the public and the private sector, as well as publicly funded research. It can also develop new tools for innovation support. Plenty of potential exists in new cooperation models between public and private sectors. For example, there is a large territory for private innovation within the owners of welfare, social and educational institutions. Coming to the scene of political decision-making may turn out to be difficult for private companies, however. Creating healthy concurrence between service providers is one of the means to increase innovation (OECD Committee for Industry and Innovation 2009).

Equally important is the role of local authorities as an indirect *enabler* of innovation. By reforming activities through regulations and standards and by removing obstacles for innovation a lot can be achieved in small and medium-sized cities. This calls for long-term engagement, cooperation, and better information for the authorities (OECD Committee for Industry and Innovation 2009). Also, the role of public authorities in financing innovation projects of the universities, research institutions, and intermediary organizations is well known. Other roles of public authorities in innovation are discussed later on.

Absorbing innovations by public authorities is by no means easy. Stiff regulation for public procurement was mentioned already. There is a European attempt to tackle this problem through precommercial procurement, which makes purchase of innovations a little easier, or at least possible. Innovative purchases often fail because of inadequate skills to work out proper criteria for selection. Shortening the time needed for procurement would also be very helpful to strengthen the role of public authorities as innovation drivers.

2 Understanding Quadruple Helix

Our main argument, stating that the civil society, citizens, should be taken into account and considered as equal players in the field of innovation, formed our initial understanding of the Quadruple Helix. The fourth pillar present in the model cannot be anything else, when talking about open, broad-based, or inclusive innovation. This is not to say that there could, or should, not be any more players—whether Quadruple Helix is enough or not is not our concern. It may well be that a *Penta Helix* is actually needed in the complex reality of the future.

The starting point of CLIQ was the baseline study CLIQboost (MacGregor et al. 2010), which outlined initial CLIQ partner profiles in view of innovation and drafted

Insight-Strategy-Stakeholder (ISS) maps to reveal the relations and the potential inherent in local innovation ecosystems of the partner regions. The study offered a wealth of information of the tools and methods that partners use in innovation promotion. It showed—among other things—that many partners have existing strong communication platforms with civil society, but what is lacking is the leveraging of these relationships for innovation. It also argued that the partners need to activate all forms of their present capital: natural, social, and knowledge capital; political leadership and continued construction of infrastructure should be consistent with the vision of better quality of life.

Interestingly enough, it made us aware of the fact that very few, if any, of the CLIQ partners had a functioning Quadruple Helix innovation ecosystem present in their regions. This was confirmed by another research commissioned by CLIQ (Arnkil et al. 2010) according to which the highest intensity actors in innovation activities in the CLIQ partner regions to date are large firms, universities, and polytechnics, national R&D institutions, science parks, and business incubators. Lowest intensity is with consumers, citizens, and employees. This manifests that there is ground for further development of the Quadruple Helix innovation system also in these regions in the future.

3 Ways to Learn

One noteworthy conclusion from the CLIQboost report was that there is a place and a common platform available for learning. Establishing relevant learning and interest groups among partners (partners with core competence matching others with learning needs) was recommended. Many interesting ways of learning both individually and collectively can be used. Important questions about learning according to the CLIQ evaluation team are: who is learning, when, what, through which methods, and with which results? "Why" is also a relevant question: to contribute to sustainable development in a complex society might be an adequate answer.

3.1 Learning from Experience

One way of learning from each others' experience is the transfer of good practices. Twenty-six case studies of good practices were published during our project. They deal with issues such as super incubation, introduction of innovative electronic tools for citizen inclusion, entrepreneurial education to all children from an early age, construction and support of local clusters, and helping SMEs to overcome times of financial crises.

Good practices were identified also during the comprehensive exchange program of CLIQ. The dozen Study Visits, Roundtables, and Master Classes carried out all had a specific theme from cluster and entrepreneurship promotion and innovation financing to the inclusion of civil society and the role of local authorities in innovation promotion. Local examples and good practices were showcased and discussed with the particular viewpoint of potential transfer from one region to another. How many of these examples and practices will actually be transferred cannot be predicted, but in principle there is nothing to make it impossible, especially, when the concept allows for modification to local needs. Economic thinking can easily recommend this kind of transfer, with recognition of the need of local customization.

One problem related to this is the valid identification of good practices. What is a good practice in general? Does it exist at all? On which criteria can we nominate a phenomenon or an action as a good practice in innovation? At the beginning of CLIQ, we did not know, and even later, the selection has not followed any unified criteria set in advance. Good practices are often local in nature and tightly embedded in the local socioeconomic context. We are inclined to think that setting common criteria for a good practice in the field of innovation is nearly a mission impossible. Even the best attempts fail because of the fact that good or the best practices are often chosen by single persons or regions, and also include an aspect of self-promotion or marketing. Even so, it is still possible that a local authority, city, or region finds new ideas, inspiration, or a missing link in one single example or practice described by a partner without too much sophistication or self-criticism, as it sometimes happened in CLIQ.

3.2 *Ongoing Evaluation*

It was a strategic choice of CLIQ to contract external evaluators to give impulse to develop the project activities during the implementation phase. The task of the evaluation team was to collect and assess especially the learning from the exchange of experience events "in real time." After each event, an enquiry was passed to the participants, and the answers were collated in a report for the project management team. According to Chelimsky (1997, 10) evaluation can fulfill various purposes: it can be pursued for accountability (e.g., measuring results or efficiency), for development (e.g., providing evaluative help to strengthen institutions), and for knowledge (e.g., obtaining a deeper understanding in some specific area or policy field). In CLIQ, evaluation was considered useful for all these purposes, but obviously for the last one in particular.

When the evaluation is supposed to enhance learning, the evaluators focus on contributing to the development but also on accumulating knowledge. In doing so, they are supposed to stimulate the participants to reflect and discuss and critically investigate the activities to gain an insight into strong and weak points. It is also the role of the evaluator to make the participants aware of what is good quality. The evaluator is no longer an inspector, but more like a teacher, a consultant, or a critical friend in a dialog with the project partners, stimulating reflection and looking at the activities from a different perspective (Karlsson Vestman 2004).

This also stimulates the participants to investigate their own and others' ideas, understanding, experiences, and concepts. It is a kind of peer review: not only the evaluator but also all the partners are active and responsible for the learning process. This emphasizes the importance of giving all the opportunity to ask questions and take part in the dialog and discussions. In this kind of evaluation, the evaluator must be present and actively participating in the events, which gives him/her the possibility to better understand and draw conclusions, to follow and analyze the learning process, and finally to decide, if the project has obtained its goal or not (Karlsson Vestman 2004).

According to the evaluation, all the CLIQ exchange events obtained their goals. You can always ask, whether the goal was the right one and correctly communicated to the participants of the event. In any case the combination of evaluator–participants and evaluation enquiries answered by all the participants should bring reliable results. For most partners, the concept of ongoing evaluation was new and met in CLIQ for the first time.

3.3 *Learning by Experiment*

One more item in the CLIQ work plan especially designed to allow for a transfer of good practices was the Pilot Project. It represents learning by doing, and eventually took a completely different course from what was planned. The experiment should have consisted of a transfer of a good practice or two from one region to the others. However, it turned out to be practically impossible to find *any* one good practice relevant to all the ten regions taking part in the pilot exercise. Therefore, a new approach was introduced, the starting point of which was the simple notion of the Quadruple Helix interaction over six relations.[1] The assumption behind this was that improvement of one part of the innovation ecosystem changes the mutual relations of the QH actors and the dynamics of interaction. It pulls along other improvements; if found useful these can be modeled and repeated.[2] Through this, the Pilot Project could improve the overall interaction of Quadruple Helix and increase the region's general capacity to innovate. The individual Pilot Cases exchanged ideas and followed up each other's progress trying to learn from the others as much as possible.

The basic question was this: How can or should the Quadruple Helix actors interact to enhance innovation in the region? Each of the Pilot Project participants were anticipated to look at the question from the specific local context and to choose one or two relations that are relevant to examine more closely. However, it was supposed that these relations are explored through a common frame, with the knowledge that at the end they will be tied together and should form a consistent whole.

[1] There are six possible relations between the Quadruple Helix actors: Administration—Knowledge Institutions, Administration—Civil Society, Administration—Business, Knowledge Institutions—Business, Knowledge Institutions—Civil Society and Business—Civil Society.

[2] The old dialectic idea of small (quantitative) steps triggering a bigger (qualitative) change.

It was also instructed to focus on connections or relations rather than on description of individual features, and to investigate the chosen focus through (already identified) good practices.

The individual cases varied from a citizen activation campaign, introduction of service design methods in a public service chain and organization of an Open Innovation Day to involving citizens in urban regeneration in a very early developmental phase. Cases were supposed to start from and embed the real needs of the participating regions and bring some added value to all of them. The outcome of the Pilot Project was expected to add to our understanding of the various methods of inclusion of the civil society in innovative action.

Some local cases included in the Pilot Project turned out to be very successful in drawing the attention of both citizens and the media. For example, in Jyväskylä having 130,000 inhabitants in a period of 2 months as much as 15,000 visits were registered in the Web site collecting and presenting ideas for the development of a former paper mill area to a residential and workplace area of the future. Using a wide variety of methods from photography to storytelling more than 600 people presented an idea or a more detailed vision for the development of the area to the city, the land owner, and the initiator of the planning phase. Many of the visions were worked out by various groups of the civil society, such as families with small children, activists in sports or culture, bicyclists, young entrepreneurs, students in secondary education, etc. The material proved to be very rich and versatile. This particular case continues online and will be included in the material for the architectural competition for planning the area. It is hard to find another example where so much positive attention would have been drawn to city planning in Jyväskylä as during the CLIQ Pilot Case.

In the Pilot Case of Brighton, UK, a Web site was developed in which the citizens entered ideas under nine categories on how to improve life in Brighton. They also voted and commented on the ideas. The top 30 ideas were presented to nine judges representing the categories of transport, leisure, health, safety, etc., and the winner was selected and awarded. The pilot was promoted heavily by the local radio and social media. A huge amount of information was gathered, partly on previously unknown problems of people's life, highlighting areas in which the citizens were most interested in.

The Pilot Project exercise contributed to the understanding of Quadruple Helix model of innovation by the CLIQ partners in several ways. First, each of the Pilot Cases addressed one or more QH relations and focused on different ones. Second, the cases were developing a QH relationship in need of improvement, perhaps where an existing relationship was not trusted or constructive as desired. Third, it wasww noticed that creating a neutral (normally online) space changes the dynamics of interaction between the "helices." In an enquiry to the Pilot Project partners conducted by Aurora Strategies and Solutions Ltd. in April 2011, it was found out, in addition, that:

- Identifying a common challenge to solve—one which engages all stakeholders—is important.
- Working with the QH improves the innovation process (100% positive response).

- All actors benefit from coworking and knowledge transfer between actors (100% positive response).
- Local authorities do (generally) have a role to play: they cannot work in isolation.
- It is important to get all QH actors involved, including senior levels of leadership to gain credibility.
- There may be negative as well as positive engagement of QH actors—you must be prepared to deal with it all.

Many of the Pilot Project participants claimed for more time to be able to include all the actors of the Quadruple Helix. Reaching out to the civil society is demanding already because of the heterogeneity of it. One approach is not good for all, and using many methods and communication channels takes more time, but may result in a very intensive exchange with a rewarding outcome.

True involvement of the civil society means also empowerment, and empowerment of a new group of actors often brings along new constraints. Public authorities may not be ready or even willing to give power to a heterogeneous group with not too much explicit expertise. Recognition of also public benefits such as better services at lower costs should, however, be influential enough to turn the heads and ears towards the citizens. Nevertheless, listening to the citizens is only the first step to be taken. Real empowerment means much more: greater well-being, better interaction, partnership—a more democratic innovation at the end.

3.4 Search for the Phantom

At some point of project implementation, our quest for the Quadruple Helix started to look quite unsuccessful: like a phantom, it escaped our grasp and comprehension. Much effort was taken to get to the core, to understand what was characteristic and vital in it. Which features need to be present, and what are the necessary and sufficient conditions of an innovation system to be called by that name? Sometimes, it looked like we were getting closer, and in the next moment a serious doubt arose: does it exist at all, is there even one single good example of Quadruple Helix to be found anywhere in the world?

What could open innovation mean in public services? It could be *crowd sourcing* (like in the case of improving the content of an article about history of a locality in a digital archive), *service design* (using methods of design to make a service more appealing or customer-friendly), or various kinds of *citizens' forums or panels* to ask the opinion of the users of public services from those who need and use them. It could also mean *cocreation* of a new service or *coplanning* of a certain area in the city, like in the CLIQ Pilot Project allowing to experiment with different methods of citizen inclusion. As explained, the main idea was to improve the connections of the Quadruple Helix actors, especially those including civil society (business—civil society, local/regional administration—civil society, or R&D institution—civil society) to enhance innovation activity and improve innovation performance of the project partners.

Even if it may be impossible to set a fixed criterion for a good innovation practice in general, within CLIQ we tried to approach the phenomenon of the Quadruple Helix through several methods, or from several points of view—to increase our understanding and the probability of finding some good examples of it. A benchmarking tool named CLIQ-o-meter was developed as an end product to measure the innovation performance. Here, we come to still another problem: exactly how do you *measure* innovation or innovation performance? Should you measure the number of innovative products (how then to decide what *is* an innovative product?), the emergence of new services (are *all* of those really innovative?), or the innovative behavior of an organization? The European Innovation Scoreboard is one answer to the problem, but we did not regard it worthwhile to make local (and much more limited) reproductions of it. This is why the CLIQ-o-meter is a self-assessment tool, by which hopefully any user can get an idea of the level of innovative behavior in the organization, whether a public authority or an innovation service provider.

Another end product, Toolkit, is designed to present some of the good practices identified in CLIQ as well as some tools used to bring them about. The examples are mostly gathered within the CLIQ network; only when there is no suitable practice or tool available in the partner regions, an example from outside the network has been used. The examples and tools are gathered under the following headlines, each describing a role that a public authority can play in innovation promotion:

- Facilitating inclusion of citizens.
- Communicating innovation.
- Supporting access to finance.
- Providing infrastructure and services.
- Developing knowledge and competence.
- Orchestrating activity.
- Formulating policy and regulation.

The CLIQ Toolkit aims to inspire both reflection and action, and it was playfully called also "CLIQ Box of Tricks" during elaboration. Each tool presented is followed by a couple of tips useful to be considered. In addition, some simple and practical top tips are presented based on the assessment of the CLIQ Team.[3]

The third end product of CLIQ, the Blueprint, tells a story of ideal conditions in which innovation can flourish. It is a vision presented in a form of an animated film that gives an idea about what we have been exploring in the 3 years of CLIQ implementation. It states that inclusion is powerful and brings benefits to all participants of the innovation process. Innovation is not only about technologies and infrastructure

[3] Alison Partridge and Sally Kneeshaw from Aurora Strategies and Solutions Ltd, London, and Tuija Hämäläinen and Laura Ahonen from the Lead Partner organization. Many ideas, tools, and descriptions presented have been developed in collaboration within the CLIQ Team.

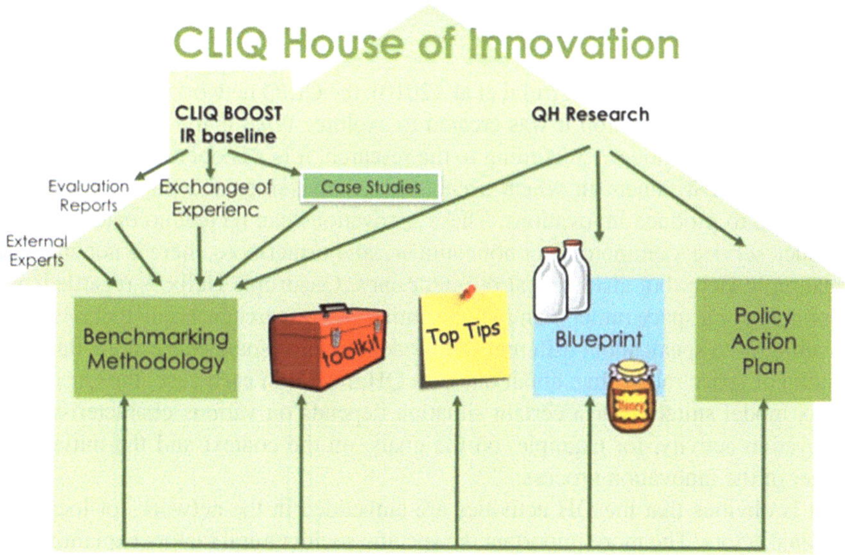

Fig. 2.2 The end products of CLIQ

but also very much about organization, services, networks, and cooperation, too. A metaphor used to describe the dream could be a finely tuned orchestra, in which the players of various instruments can play alone or all together, depending on the composition. Even if the players can play independently, the conductor gives the performance his/her unique touch, the final refinement. The impression of the audience is the result of the whole orchestra, each and every player who contributes to the performance trying to follow the gestures of the Maestro as truly and precisely as possible. At the end, we decided to use another metaphor, however.

Finally, a Policy Action Plan paves the way forward to the CLIQ partners as well as to other public authorities and innovation service providers in medium-sized towns trying to improve their performance in driving innovation. By setting the policy context on the macro level and identifying relevant thematic areas, the Plan outlines through examples and general strategic considerations policy in action in the partner regions. It even goes a step further and suggests a set of actions to be adopted by the partners. Designing specific multi-annual local/regional innovation strategies and implementing them in a comprehensive collaboration with stakeholders and steered at the highest political level is one of the suggested actions. Creating networks of learning as well as subnetworks of universities, science parks, and incubators with other cities and regions having the same aim is another. Orchestration of the common effort to make a change happen can be mentioned as a third example.

In summary, CLIQ will make available the collected results of 3 years of learning in these specific end products, which could be seen as constructing a House of Innovation (Fig. 2.2).

4 Quadruple Helix Revisited

Along with the research by Arnkil et al. (2010), the CLIQ network found a theoretical answer to the question it was created to explore: What is the Quadruple Helix (QH) innovation model? According to the research, it is a cooperation model of an innovation environment in which users, firms, universities and public authorities cooperate to produce innovations. These innovations can be technological, social, product, service, commercial, or noncommercial. Furthermore, there is not only one Quadruple Helix but also several different ones. Quadruple Helix is regarded as a continuum or a space rather than a single entity, and researchers argue that it is more meaningful to speak about different QH models situated somewhere along this continuum or space rather than about one best QH model. In each case, the Quadruple Helix model suitable for a certain situation depends on various characteristics of innovation activity, for example, on the goals, on the context and the initiator or owner of the innovation process.

It is obvious that the QH activities are embedded in the networks of local and regional actors. The more important the specific sociocultural factors explaining the formation and development of a region are, the more difficult it is to transfer experiences from one region to another. The civil society is mainly a local or territorial phenomenon, with some global dimensions though. It is essential for local and regional authorities to learn that there are different useful models for promoting civil society engagement in innovation, and that their capacity to include citizens may vary. Also, the role offered to public authorities within the four basic models distinguished is different. A broad mix of concepts in use will most likely bring the best outcome.

The QH research outlines four different types of QH models with various degrees of citizen/user[4] involvement: (1) Triple Helix + users, where the traditional Triple Helix is enlarged by citizens or users who give information about their needs and experiences, typically for example testing products or services at a late developmental phase, (2) the firm-centered Living Lab model, where the citizen or user, in addition, participates in the idea and development phase of an innovation, but business remains the main driving force, (3) the public-sector-centered Living Lab model, which compares to the previous model with the difference of having public authorities in the central position, and (4) the citizen-centered model, where finally the user decides which innovations are needed and developed and where the citizen is really in the center of the cooperation platform. These are seen as ideal types, not existing in one-to-one relationship to reality. The development is not linear (from a lower to higher degree of citizen engagement), and there is probably a mixture of some or all of these models existing in and available for the regions. Different roles played by public authorities in the Quadruple Helix type of innovation identified by the research are those of an Enabler, a Supporter, a Decision-Maker, a Utilizer, a Developer, a Marketer, and a Quality Controller (Arnkil et al. 2010).

[4] Here, the civil society means broadly understood users (consumers) who are using the products and services produced by firms and services produced by public organizations (Arnkil et al. 2010).

According to the QH research, local and regional authorities have an important role in the Quadruple Helix via strategic use of resources, integrating knowledge and skills in innovative thinking, community building, procurement, regulation, grants and rewards. In order to succeed in this, the authorities need to develop their own ability and skills, and to cope with constraints, inflexibilities, and the bureaucracy inherent in public organizations. They are faced with the challenge of renewing themselves to be an interesting partner in reforming the local–regional innovation ecosystems. All the aspects of demand and user driven innovation policy—development of skills, reforming regulation and the operational models of the public sector, and introduction of incentives—need to be considered to build an innovation ecosystem that could be characterized as Quadruple Helix.

No matter which method of citizen involvement is used, it necessitates a strong communication effort. Taking it for granted that a majority of people are interested in their living environment and the services they use on a daily basis, their interest needs to be turned into action, and drawing from the CLIQ experience, this conversion does not happen very easily. The opportunity needs to be communicated again and again, many times both directly and indirectly, using all the communication channels available. The threshold from seeing a problem to addressing it is often higher than one could imagine. Complaining about bad public services is a kind of common entertainment, but taking actual steps to improve the services is by no means self-evident and needs to be encouraged, enabled, and enacted through a variety of measures, not least through effective communication.

There are successful communication actions to be reported in CLIQ in many partner regions. We have come a long way since the beginning, when we hardly understood what a Quadruple Helix could mean in practical terms. Questions have turned into statements and these have become bases of action that we could never anticipate. This is part of the charm, but also part of the challenge of an experimental innovation project like CLIQ.

How does the Quadruple Helix model of innovation look after this specific practical experience? What is the collected learning from CLIQ? We summarize as follows:

1. There is a real need to improve and enlarge the concept of Triple Helix, which can no more incorporate new thinking and new concepts targeted to enhancing innovation activity.
2. There are many ways of doing so, not only one answer or one good solution suitable for all. Quadruple Helix can mean different things in different contexts; only the imagination sets the limits. "It is more of a continuum or space than a fixed concept," stated the research commissioned by CLIQ.
3. No matter which method you use, it must be communicated effectively and extensively to include and motivate all the target groups. Motivation takes time, but as soon as born, it may bring about astonishing results. Remain optimistic, creative and ready to talk it over and over again—and you will succeed.
4. Getting relevant proposals or solutions may be slower and more difficult than expected. Design special methods of eliminating spam, as well as ways to award the best proposals.

5. The results may be something unexpected and radically new. They may also be only one small improvement in the design or in a complex set of services. It is impossible to anticipate due to the very nature of open innovation: it is open until the end.
6. Quadruple Helix can allow more inclusive and even open innovation, thus bringing benefits that other innovation systems are not capable of bringing about. Inclusion of the civil society—citizens—in innovation is vital; it is particularly important to develop the efficiency and transparency of the public services.
7. Creating an operational Quadruple Helix innovation ecosystem is challenging for many reasons, not least because of the need to change the working methods and develop new abilities and skills. Empowering user groups—that is, citizens—changes also the roles of the players and the rules of the play, which may not always be easily accepted or appreciated.
8. More inclusive innovation means better cooperation and more comprehensive networks to be exploited in collaboration. Developing these (local, regional, and interregional) networks is an investment of time and effort, but it will pay off. As stated by Magnin (2010, 10), a territory ready to network with its outside world is best equipped to tackle the energy and climate challenges we are facing today while giving priority to the quality of life of its inhabitants.

There is no one success formula for the development of a more innovative region. Quite often, the local authorities, who in fact can influence innovation promotion in their daily life actions, are not aware of this, because innovation enhancement is a new aspect in their work. During the CLIQ project, we have tried to improve the effectiveness of policies and instruments for regional development by exploring, optimizing, documenting, and promoting the role of local authorities so that their efforts would be fruitful and could lead to more and enhanced innovation.

However, a region or an organization does not become successful by itself or by coincidence. It is obvious that the success story of the city of Ulm, Germany, as an example, is due to a long-term commitment, shared visions and responsibility, and collaboration between all the four "helices" (certainly not referred to as such when it all started). The importance of a strong and committed leadership combined with an openness to involve groups and individuals allowing them to influence, but also to take on the responsibility for the processes, was manifested during the CLIQ study visit in Ulm. The collaboration between the four helices, strategically, scientifically, and economically, in the field of sustainability (theme of the visit), was according to the evaluation the outstanding issue,[5] easily recognized by the partners, too.

During the project, we have learned that the cities the CLIQ partners represent have many strengths, such as flexibility and good cooperation, but they often seem to be less well equipped in terms of critical mass, resources, and organizing capacity. Learning from the others' experience is cheap compared to learning by mistake. An interregional project such as CLIQ gives its participants many opportunities to reflect on one's behavior and compare it to that of the others. It does not matter

[5] Falk M (2011) Evaluation of Ulm. Mälardalen University, Eskilstuna.

where the impulse to improve comes from; what matters is the improvement itself. It is vital to keeping our economies competitive, as emphasized also by the Innovation Union initiative of the European Union. Smart, sustainable, and inclusive growth resulting from smart, sustainable, and inclusive innovation is the key to the economic success of Europe in the future.

The Interreg program is about improving regional policies. We have tried to collect some seeds for thought through various activities in CLIQ. Finally, it is up to the political decision-makers, if they pick these seeds up and let them grow. All local and regional stakeholders are needed to cross-fertilize ideas and to support the initial growth. It is good to bear in mind that not all the seeds will sprout, and not all our partner regions can become innovation hotspots of Europe. To be recognized as such, two things are crucial: continuous learning and visionary leadership. With these provided, there is a fairly good chance for sustained success.

References

Arnkil R, Järvensivu A, Koski P and Piirainen T (2010) Exploring Quadruple Helix – Outlining user-oriented innovation models. University of Tampere, Institute for Social Research, Work Research Centre, Tampere

Boivie Y, Falk M and Tegerstrand C: evaluation reports of CLIQ 2009–2011. Available at http://www.cliqproject.eu

Browne L, Poikola A, Ruodemäki A (2011) Kaupungin Kangas, pilot of participatory city planning in Jyväskylä (available at www.jyvaskyla.fi/kangas)

Chelimsky E (1997) The Coming Transformations in Evaluation. In: Chelimsky E and Shadish W R (ed) Evaluation for the 21st Century. A Handbook. SAGE Publications Inc, Thousand Oaks, California

Communication of the European Commission (2010) Europe 2020. Strategy for Smart, Sustainable and Inclusive Growth. Commission of the European Union, Brussels

Interreg IVC Innovation & Environment. Regions of Europe sharing solutions (2010). Programme Manual. Available at http://i4c.eu/resources_programme_manual.pdf

Karlsson Vestman O (2004) Evaluation as learning. Research report 2004: 8A. CVF&MEA Mälardalen University, Västerås & Eskilstuna

Kysyntä-ja käyttäjälähtöinen innovaatiopolitiikka – Jäsentely (osa I) ja toimenpideohjelma (osa II)

Title translated: Demand and user driven innovation policy – Analysis (part I) and Action Plan (part II). Publication No. 47/2010 of the Ministry of Employment and the Economy of Finland, Helsinki

Magnin G (2010) Low energy cities with a high quality of life for all. Energy Cities. Available at http://fr.calameo.com/read/0001260423b56a6f28003

MacGregor S P, Marquès P, Simon A, Bikfalvi A, Llach J (2010) CLIQboost, baseline research for CLIQ Interreg IVC. Documenta Universitaria, July. ISBN: 978-84-92707-34-8

Kansallinen innovaatiostrategia (2009). Title translated: National Innovation Strategy. In: Innovation Policy Lines 2010. Ministry of Employment and the Economy of Finland, Helsinki

OECD Committee for Industry, Innovation, and Entrepreneurship (2009) New Nature of Innovation. OECD Committee for Industry, Innovation, and Entrepreneurship, Copenhagen

Reports work papers and presentations by the CLIQ Team: Kneeshaw S and Partridge A from Aurora Strategies and Solutions Ltd, London together with Ahonen L and Hämäläinen T, City of Jyväskylä, Administrative Centre. Available at http://www.cliqproject.eu

Presentations and evaluation reports by the CLIQ Evaluation Team: Ylva Boivie, Clas Tegerstrand and Monica Falk, Mälardalen University, Eskilstuna. Available at http://www.cliqproject.eu

Chapter 3
The First Bank of Cents: Innovative Carpooling Through Social Currencies

Josep Lluis de la Rosa and Andrea Bikfalvi

Abstract The First Bank of Cents (FBC) provides social currencies, a type of money for social usages, to feed the behavior shift of targeted communities into more sustainable development. FBC can turn driving into a social asset by reinventing the carpooling practice through rewards for social drivers, those who share their cars, and social networks that provide trust among users and confidence on having safe round-trip transportation. FBC provides the rewards so that social drivers obtain cheaper fuel, car maintenance, and parking; hitchhikers comfortably, and with trust, ride for free most of the time; and carpooling works as a powerful loyalty marketing service for local, sustainable businesses. FBC shows how different entities (business, academia, government, and society) can add value through stronger networks that better involve and engage different actors—companies, customers, and knowledge generators, among others. FBC could therefore be viewed as putting the Quadruple Helix into practice, as it seeks to engage public administrations, users, and private enterprises around a common cause of sustainable mobility, which also has the added bonus of being socially responsible.

J.L. de la Rosa (✉)
TECNIO – Centre EASY, University of Girona, Parc Científic de Girona,
Girona E17071, Spain
e-mail: peplluis@eia.udg.edu

A. Bikfalvi
Department of Business Administration and Product Design, University of Girona,
Av. Lluís Santalo s/n, Escola Politècnica Superior, Edifici PI, Girona E17071, Spain
e-mail: andrea.bikfalvi@udg.edu

S.P. MacGregor and T. Carleton (eds.), *Sustaining Innovation: Collaboration Models for a Complex World*, Innovation, Technology, and Knowledge Management,
DOI 10.1007/978-1-4614-2077-4_3, © Springer Science+Business Media, LLC 2012

1 The Quadruple Helix in Practice: Academia, Firms, Government, and Civil Society

Economic growth is, undoubtedly, generated by concentrated human talent and knowledge. According to the Quadruple Helix model (4H), a country's or region's economic structure lies on four pillars: academia, business, government, and the civil society, each of them having specific roles and objectives. Although specific and differentiated, complementary relations unite them for the general purpose of creating a knowledge-based society characterized by socioeconomic welfare.

The turn of the century marks an important *before* and *after* in terms of innovation. According to Afonso et al. (2010), before 2000, innovation was often mapped by using the national systems of innovation approach, a basis for arguing how different organizations and institutions contribute individually or collectively to generate innovations. In the 2000s, a series of changes in views of innovation systems took place resulting in a series of novel approaches, such as global networking and value added for innovation, system thinking and innovation ecosystems, customers and users in the innovation process, to mention just a few.

Moreover, new settings for innovation have emerged resulting in specific typologies or labels for innovation that go beyond the already classic categories—product, process, incremental, radical, etc.—described by traditional innovation manuals. Some examples refer to *eco* innovation, *open* innovation, *sustainable* innovation, *user-centric, focused* or *driven* innovation, and *living labs*.

Figure 3.1 shows a simplistic, but illustrative, evolution of innovation models over time. It is worth highlighting the increasing role of innovation networks and knowledge clusters for knowledge creation, diffusion, and use.

According to Carayannis and Campbell (2009):

> [A system formed by innovation networks and knowledge clusters]…is a multi-layered, multi-model, multi-nodal and multi-lateral system, encompassing mutually complementary and reinforcing innovation networks and knowledge clusters consisting of human and intellectual capital, shaped by social capital and underpinned by financial capital.

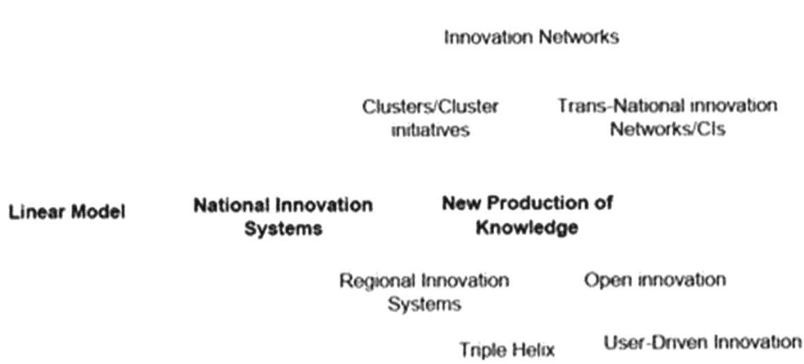

Fig. 3.1 Evolution of innovation systems (Arnkil et al. 2010)

It goes beyond the purpose of the present chapter to analyze the creation and the evolution of the Quadruple Helix (4H) model, or to further illustrate trends in mapping the innovation system. This section rather focuses on an illustrative example of applying the 4H model to the concrete case of an electronic carpooling system—presented from FBC's perspective—describing each participant's role, implication, and perspective. It is therefore our vision of sustainable innovation as applied to mobility within the general context of sustainable development.

1.1 Government Perspective

The EU states: "Sustainable Development stands for meeting the needs of present generations without jeopardizing the ability of futures generations to meet their own needs – in other words, a better quality of life for everyone, now and for generations to come. It offers a vision of progress that integrates immediate and longer-term objectives, local and global action, and regards social, economic, and environmental issues as inseparable and interdependent components of human progress".[1] From a strategic point of view, there are seven key priority areas: climate change and clean energy, sustainable transport, sustainable consumption and production, conservation and management of natural resources, public health social inclusion, demography and migration, global poverty and sustainable development challenges.

At a more local, regional, or municipal level, sustainable development is in each governmental body's discourse irrespective of its political color. Municipalities and local entities cowork to manage the change in thinking, in economic and social structures regarding production and consumption. Concrete initiatives target especially the environmental facet of the issue that later might turn into social and/or economic benefits.

FBC's initiative concerns one of the seven mentioned key challenges, namely, sustainable transport. Ideally, the specific roles that the local administration could assume would be the following:

- Facilitate the use of the system.
- Create awareness on its existence, functioning, and impact.
- Be a first user and generate a role model acting as an example.
- Provide financial support (grants, etc.).

At a time when local authorities search for arguments and concrete solutions for specific problems, complying with their general role of creating an environment conductive to innovation, their involvement in the FBC initiative is fundamental. Their position is double: first they intervene as a promoter, and second they can assume a user role also.

[1] http://ec.europa.eu/environment/eussd/ (Last accessed March 8, 2011).

1.2 Business Perspective

In the business domain, sustainable development issues have also accelerated in the past decade. They are most often known under the Corporate Social Responsibility (CSR) umbrella term. It is the company response to the societal wide scope of sustainable development, addressing at its core the relationship that the business has with society and the "social contract" it possesses with the public. According to the most recent *Guide to CSR in Europe 2010*[2] recent trends show that growing attention is paid to the voluntary actions that companies take as part of their CSR strategies to manage their economic, social, and environmental impacts and to contribute to wider societal development.

Recently, increasing interest in business opportunities associated with innovative CSR approaches is an issue to consider. Furthermore, as a result of the financial and economic crisis, the level of public trust in business has fallen in many European countries; therefore, it is crucial for companies of all sizes to contribute to rebuilding trust in business and shaping a more responsible and sustainable economy in Europe and globally.

There are several important lessons for developing a successful carpooling initiative. On the one hand, the wider stakeholder model fits with the general model of FBC, necessarily including both direct users but also other partners who are able to influence uptake and also provide motivation through the use of complementary currencies. An analysis of relevant stakeholders and the value for them as well as the claims they may make on a system is detailed later in this document.

CSR may also offer an opportunity to gain access to markets, especially large companies who are looking to differentiate their actions, and improve on their reputation. Although the more important aspects of CSR should not be solely based on the marketing view as noted above, this remains the focal point of much current perception.

Carpooling is a small, yet growing element of company's CSR development and the budgets they have to implement policy are growing. Companies should move beyond anecdotal reports on CSR and implement apparently simple, but effective solutions that benefit the community. FBC aligns well with such a purpose.

1.3 Academic Perspective

Academia is a key player in knowledge/novelty generation. It provides knowledge for companies that further develop and commercialize them in forms of innovation. It is the main human resource trainer and provider for the society in general. Universities' role in the knowledge society has changed over time, passing from education to then education and research and nowadays to education, research, and research commercialization (EC, 2003).

---

[2] http://www.csreurope.org/data/files/2342aguidetocsr02.pdf (Last accessed March 8, 2011).

Higher education in Europe is facing an imperative need to adapt and adjust to a series of changes, such as increased demand for higher education, the internationalization of education and research, to develop effective and close cooperation between universities and industry, the proliferation of places where knowledge is produced, the reorganization of knowledge, and the emergence of new expectations. It is in this context where the dialog between the university and the society should be improved through a variety of concrete actions: work in a long term perspective, communicate the key role of research in underpinning university autonomy, consolidate links to different stakeholders and strengthen links at local/regional level where the importance of universities in the life of their communities is growing rapidly (EUA, 2011).[3]

Innovation literature, both scientific and nonscientific, is abundant in case studies describing the successful path an idea from a university took to reach a market. Entrepreneurial initiatives from the academic community are valuable in a variety of senses: they represent the practical application of the knowledge created; they offer professional future for young students; and they increase the reputation and trust of the academic founder.

The social entrepreneur movement has shown that sustainable development through its social and environmental facets is a field where business opportunities arise, in the sense that individuals create and exploit business value at the same time as solving a social and/or environmental problem.

From the perspective of FBC, academia plays a double role. On the one hand, it provides the environment conductive to the idea of the project, the majority of the personnel involved, and knowledge necessary to the idea's development and application. On the other hand, academia is a powerful user since it can be considered a community located in a rather delimited space employing a variety of collectives (e.g., faculty, students, etc.) that might find travel sharing commonalities. Of course, the creation of FBC may be viewed as a social entrepreneurship initiative.

1.4 Society Perspective

Awareness of environmental and social issues has increased dramatically over the past decade, as the world questions the future of the planet. Civil society contributes—both positively and negatively—to shape this future. Individuals need specific reward systems (as opposed to punishment) to take up and consciously use and apply certain initiatives. A perceived value and a positive impact should be main triggers in implementing social and/or environmental solutions. Previous positive experiences or similar projects serve as role models or illustrators for them.

Going beyond the individual and considering the macro level the wider societal viewpoint is worthy to analyze. The sum of individuals/users and the relationships established among them result in communities that can be motivated and incentivized

[3] http://europa.eu/legislation_summaries/education_training_youth/lifelong_learning/c11067_en.htm (Last accessed May 9, 2011).

toward contributing to sustainable development. The key issue remains motivation and trust. The FBC initiative should make clear that users and their corresponsive community and action—by participating—contribute to reduce the environmental damage to the environment through reduced emissions.

1.5 Integrating the Quadruple-Helix

In the previous sections, we presented the role and contribution of each involved actor from a 4H perspective. It is interesting to highlight that roles should be clear because all institutions—government, business, academia, and society—might assume a variety of roles. The FBC initiative is an illustrative case study on how the different 4H pillars integrate and complement in a success-oriented social entrepreneurship project. All actors involved are united by the umbrella term of sustainable development. The user–promoter double role applies for each integrator and specific motivators should be designed toward an effective use. Traditional innovation typologies often suppose important investment. Nowadays, alternative or complementary innovation typologies could supplement companies' strategic choices. We refer to social innovation together with service innovation as possible candidates.

2 The Vision and Mission of FBC

The First Bank of Cents (FBC) is a service for providing riders and drivers with enough social currency for their shared short trips, for developing marketing programs for local development, and for supporting the deployment of sustainable mobility. Its business model is twofold: first, exchange and trade commission among riders, drivers, and businesses, and second, sales and consultancy of a software platform that deploys sustainable mobility.

FBC shows how different entities can add value through stronger networks than better involve and engage different actors—companies, customers, and knowledge generators. It could therefore be viewed as putting the 4H into practice, as it seeks to engage public administrations, users, and private enterprise around a common cause, which also has the added bonus of being socially responsible.

Money is very important for our society and much more in today's tough economic situation. There are social movements, such as sustainable mobility, which not only helps to improve the environment or the traffic in a community, but also helps to stimulate the economy and the commerce in a region or community through carpooling and economic exchanges that are generated through the realization of shared trips.

The initial conception of FBC was through the e-Hitchhiking real-time carpooling platform (http://www.e-hitchhiking.com) that had a focus on car owners and public transport users in dense urban areas—those who are highly motivated to find alternative, convenient travel solutions at low cost. The actor network for FBC in this application is presented below in Fig. 3.2, with a dual social currency of "Dits" and "Fets" shown in operation in Fig. 3.3.

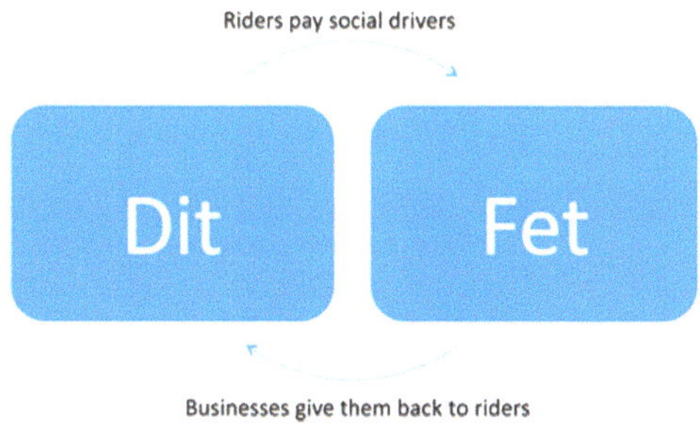

Fig. 3.2 FBC applied to the e-Hitchhiking system

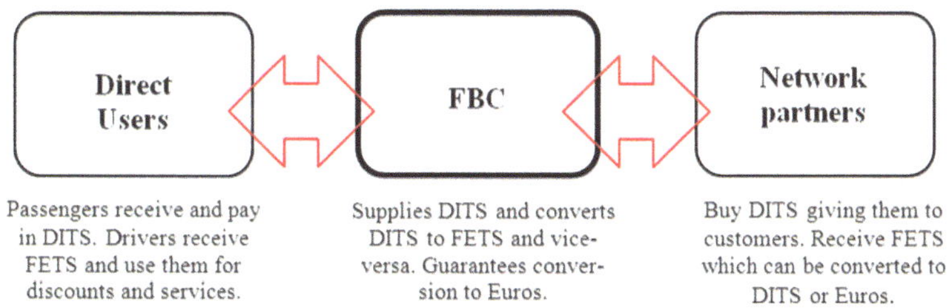

Fig. 3.3 Dual social currency of "Dit" for a ride request and "Fet" as the acknowledged ride by drivers

The meaning of "Dit" is "thumbs up" in Catalan, and the gesture represents a free ride for a passenger. The term tries to catch the idea that every time you show your intention to ride for free in any other car, you show your thumb in the street. The number of Dits corresponds to the number of free shared rides. The meaning of "Fet" is "fact" in Catalan, and it represents that the driver has truly delivered social benefit through sharing their car. Thus, there is a flow of Dits toward Fets (thumbs-up toward facts) from riders to drivers. Accumulated Fets are exchangeable for a series of rewards from FBC partners.

This is the way FBC is applied to sustainable mobility: it creates a network or community around a specific application, allowing different user types (ultimately FBC customers) to satisfy their different needs by constructing win–win relationships based on the use of a complementary currency. Among other revenue generation modes, FBC receives a commission on the transactions that take place between actors in the network.

The flow of the social currency is as follows (noted in Fig. 3.4). Companies buy Dits with Euros (rate of 1 Eur = 1 Dit) and distribute them to people, who will use them as a means of payment for free rides. Drivers receive Fets as reward (namely, 1 Fet per every traveler that had shared one ride in their cars) and accumulate Fets until they decide to use them for variable discounts in fuel, car maintenance, and other benefits. Those companies granting the variable discounts convert the Fets into Euros. FBC charges a commission after every trade. The exchange rate is of 1 Dit equals 1 Fet equals 1 Euro, guaranteed at all times by FBC.

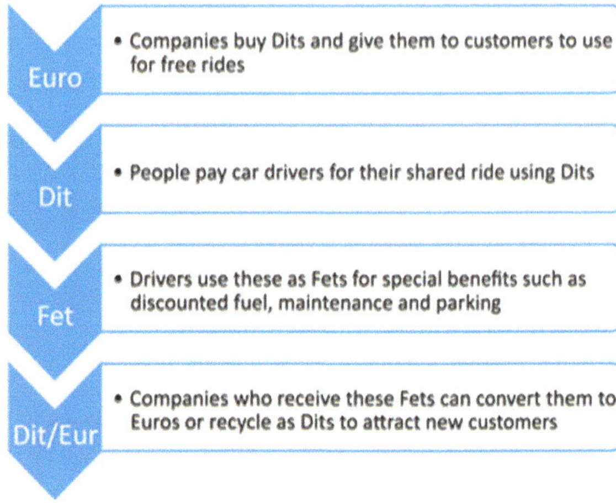

Fig. 3.4 Euro and Dit and Fet conversion by FBC

Users are able to move from place to place at a far reduced cost, which will come into sharper focus within the coming age of austerity[4] and spiraling energy costs. Social pressures are increasing affecting citizen behavior in line with sustainable development and environmental care, while city center mobility is an increasing problem with rising populations worldwide.

FBC is uniquely positioned to exploit this opportunity given its strengths in technological development and academic rigor. The e-Hitchhiking technological platform has already been the focus of a 12-month pilot project in Girona Science and Technology Park, and key agreements are in place for the first live implementation for the City of Girona during 2011. FBC also defends the idea through its strategy of collaboration to compete. It does not plan to meet "head-on" comparable products on the international stage, rather to help various applications with providing incentives and achieving sufficient critical mass through the transfer of a franchise based model that helps to create unique networks.

With reference to the e-Hitchhiking system, a summary of customer/user types and their respective needs is detailed below. We also include user scenarios for each type of customer (or network actor).

Carpooling passengers: Direct users of the system who are looking to get from one place to another by hitching a ride with the car owner. The system can facilitate routine journeys although the real value is in the dynamic matching of journeys via GPS.

USER SCENARIO: Joan does not own a car, and he normally takes the bus to go to the university. He enjoys sports and has lunch each day at the restaurant of the science park. At the end of the month, he receives a total of 20 DITS from his gymnasium and the restaurant, which he then uses to share 20 car journeys with his friend Pol the next month.

[4] http://www.telegraph.co.uk/finance/personalfinance/savings/7704990/Prepare-for-the-age-of-austerity.html (Last accessed March 15, 2011).

Carpooling drivers: Car owners who share their car with one or more passengers for an agreed journey. As with the passenger case, journeys may be matched in real time using GPS.

USER SCENARIO: Pol normally drives to the university and has received a total of 180 FETS from Joan, Marc and Laia, with whom he has shared his car. He uses these FETS for a discount in his local petrol station, a preferred parking space in the Barri Vell campus which has limited parking, and one shared car journey as a passenger.

Public entities: This segment includes all public administrations who wish to achieve policy objectives related to sustainable mobility and quality of life.

USER SCENARIO: Girona City Hall is a partner in the e-Hitchhiking system at the university and has received 500 FETS from drivers this month requesting access to the reserved carpooling parking spaces. It has decided to convert these FETS into DITS, as well as buying 200 more DITS from the First Bank of Cents to give out to its employees and help reduce city center congestion, pollution, and noise.

Other promoters: These include large localized sites who wish to address problems related to mobility and accessibility, such as industrial estates and science parks, universities, hospitals, airports, and large enterprises (e.g., the 12,000 worker site of Telefonica's headquarters [Distrito C] in Madrid). They encourage use, perhaps by offering privileges within the scope of their control, and the system is able to easily attain a localized critical mass. Regarding the multimodal use of public transport within the context of sustainable mobility (e.g., combining bus and metro to reach a destination), public transport companies may also be involved in promotion.

USER SCENARIO 1: This month, the University of Girona has bought 1,500 DITS from the First Bank of Cents to give out to its students and staff. It has also granted 75 preferential parking spaces to students and staff for the value of 1,250 FETS, and has decided to convert these to 250 DITS to give out next month.

USER SCENARIO 2: In the same month, the local public transport company has received 400 FETS from passengers and given out 900 DITS to frequent travelers. They decide that for next month, they will exchange the 400 FETS received and buy 500 DITS to support their incentive scheme to travel more often by bus.

Car-related products and services: This network type includes petrol stations, car garages, and insurance providers whose target market is car owners and benefit from exclusivity agreements with a target population. In the start-up phase we look especially to mid-range or growing companies, not sector leaders, who are looking for avenues to disrupt the incumbent leader.

USER SCENARIO: The local petrol station GironaOil has this month supplied petrol to the value of 550 FETS and has decided to exchange 300 of these into DITS for next month to give to their clients to encourage custom, and exchange the remainder into Euros.

Other commerce: This includes all other commercial interests unrelated to the use of the car, principally bars and restaurants. These users therefore fill unused capacity and can target specific users to fulfill their needs. A further benefit regards the scale of the commercial venture—local commerce may be interested in simply fulfilling capacity, whereas larger, multisite enterprises and franchises may also be interested in the market intelligence provided by the system (demographic and mobility information). The power of such a system can be viewed from the US-based NuRide, which has annual turnover in excess of $1 M uniquely through a commission based model on creating networks of consumption.

USER SCENARIO: The Girona city center restaurant, El Platano, has suffered from a sharp decline in midweek dining as the effects of the economic crisis take hold on their normal clientele. They sign an agreement with FBC, and 3,000 users of the e-Hitchhiking system in the university are entitled to special meal offers during the week.

FBC can therefore be viewed as *exploiting the latent power of web 2.0, creating, consolidating, and growing communities, and monetizing* their relationships through the use of a social currency. The proven case of loyalty programs is the main lever and translates to discounted goods and services on the demand side and increased consumption on the supply side. Public level administrations are also able to achieve desirable social behaviors. FBC is driven by the use of social currencies.

3 Social Currencies, a Type of Money

The FBC vision is to provide social currency, as a new type of money for social usages, to feed the shift of behavior of targeted communities toward sustainable development. We see one example: the sustainable mobility by means of intermodal carpooling.

Money is our oldest information system as "writing" was invented in Mesopotamia as a method of "book-keeping" (Lietaer 2001). The earliest texts available, from 3200 BC in Uruk, are record of various financial transactions, including secured and unsecured lending and "foreign exchange" transactions. Money is our most pervasive information system, as it percolates through billions of daily exchanges in all strata of society. Today, it has become a truly global information system—now that trillions of dollars are moving at the speed of light in a totally integrated, round-the-clock, computerized foreign exchange market.

Social currencies are an alternative currency to legal money. The rationale for social currencies is this (Seyfang and Smith 2007): since money is a socially constructed institution, it builds in certain characteristics and carries with it specific incentives and inherent values to promote particular types of behavior. This approach is in contrast to the conventional economists' view of money as the neutral technology, which oils the wheels of economy activity. Taking "a whole systems" approach to the socially—and environmentally—embedded context of economic activity, new monetary systems can be designed to prioritize different behavior patterns, to target different group of users, and to incorporate different "essential functions" of money (normally considered to be: unit of account, store of value, and medium of exchange). For instance, they may reward labor which is normally not valued, or they may shift the balance of market signals to favor local production over imports.

The social currency movement emerged in a number of countries during the 1980s and 1990s. Two indications of its extent are the establishment in 1997 of the online, peer-reviewed *International Journal of Community Currency Research* (Krohn and Snyder 2008) and the older working exponent of the WIR of Switzerland, in continuous operation from 1936 (Stodder 2009). Additionally, the Complementary Currency Resource Center[5] was launched in 2004 to collect statistics on a broad variety of indicators related to the design and function of all types of social currency systems in use. The impact and benefits of social currencies on the economy has been discussed by several authors. Krohn and Snyder (2008) state that local currencies have a history of spontaneously arising to the benefit of local populations in circumstances of inadequate banking services, shortages of money, and high unemployment like those of Argentina in the early 2000s.

Social Currencies have also been designed and implemented to face environmental, social, and economic aspects of sustainable development. Seyfang (2009) and Blanc (2011) have grouped them as follows:

- *Local Money Systems (LETS)*: LETS combines social and economic objectives, but principally operates through a parallel economy designed to strengthen local economic linkages. Members of a LETS exchange goods and services without using cash, using local transactions, recording credits and debits with the system accountant. No interest is charged or paid, so there is no incentive to hoard credits, and facilitating local exchange becomes the primary objective.
- *Time currencies (Time Banks)*: Time banks aim to build social capital through supportive community networks, and institutionalize reciprocal self-help through a central broker. Each hour of a service—e.g., dog-walking, a lift to the shops—is worth exactly the same, onetime credit, and participants earn credits by helping others, and spend credits receiving help themselves.
- *Green Reward Points (NU)*: It is a "green loyalty point" which was piloted in the city of Rotterdam in the Netherlands from 2002 to 2003. It functions as a reward card, similar to supermarket loyalty points, and targets environmentally friendly consumer behavior, so providing incentives to switch consumption patterns. Green points are earned when city residents separate their waste for recycling, use public transport or shop locally and they can be redeemed for public transport tickets or discounts on sustainable products.
- *Personal Carbon Trading (PCT)*: The Personal Carbon Trading idea was developed in 1966 and envisaged progressively stricter rationing to be the only plausible method of achieving large-scale cuts in carbon emissions. The benefits of PCT over regulation and taxation are that it allows flexibility of response and engages a sense of common purpose and active citizenship.

The fact is that money changes societies. And for FBC we intend to use money as social currencies, to turn driving into a social asset, but what of the conventional barriers associated with carpooling? We discuss this challenge in the next section.

---

[5] http://www.complementarycurrency.org/ccDatabase/les_public.html.

4 Background on Carpooling

Carpooling consists of sharing a single car with other users who make similar journeys. The benefits are multiple: on a personal level, it saves fuel costs; on a social level, it helps reduce traffic congestion and carbon dioxide emissions. Carpooling is considered one of the principal alternatives within the emerging global movement toward more sustainable mobility policy.

The increase of traffic in urban areas has lead to a situation of "chronic congestion" in many cities across Europe. This has several negative effects in terms of daily life and environmental impacts. According to the EU Green Book "Towards a new culture for urban mobility" [COM (2007) 551], the European economy has around 100,000 M Euros of annual losses due to mobility problems in urban areas. The Green Book mentions five challenges to face: free-flowing towns and cities, greener towns and cities, smarter urban transport, accessible urban transport, and safe and secure urban transport. "Carpooling" and control intelligent systems are mentioned as relevant options to work toward free-flowing towns and cities.

In spite of the advantages of carpooling, there has been a downward trend in vehicle occupancy in developed countries in the last 30 years. In the UK and in the USA, the average car occupancy has stabilized at 1.6 persons per vehicle and for commuting journeys, even though most cars have four seats (Santosa et al. 2010). The reasons for this decline have been analyzed by several authors[6] (Ferguson 1997; Levin 1982):

- Comfort and convenience, and increased time commitments.
- Decreasing fuel costs: carpooling reached its peak during the oil crisis in the 1970s and has fallen down since then.
- Improved vehicle efficiency.
- Smaller families.
- Higher demands for greater privacy.
- Lack of flexibility: when trips must be planned in advance the system becomes less attractive. The "plan-free" travel concept only can be ensured through real time data availability and processing.
- Reliability of users: the system must incorporate semi-automated tools to monitor when a user is to be trusted and when not. A driver must be sure that travelers who get into his/her cars are reliable and vice versa.
- Motivation of users: the system should incorporate mechanisms to motivate users, especially drivers, to use the system.
- Lack of critical mass: to achieve a system that attracts riders there must be many drivers available. Potential passengers (or riders) must know that there is a robust bourse of vehicles available at any time.
- Legal and regulatory constraints. Payment of rides to drivers through an electronic platform is not allowed in some countries like Spain.

[6] http://dynamicridesharing.org.

The potential of carpooling in relation to fuel and emissions savings has been analyzed by Jacobson and King (2009) and Lindqvist and Tegner (1998). Theoretically, the potential for fuel savings in the USA from increased ridesharing in noncommercial passenger highway vehicles is substantial. However, this benefit is offset by the need to travel additional distances to pick up passengers, and many drivers may view the time spent in picking up passengers as more costly than the fuel savings. However, if no additional travel were required, the effect of adding one additional passenger in every 100 vehicles would lead to an annual savings of 0.80–0.82 billion gallons of gasoline. Travel trends indicate that the value of time for travelers must be below $4.24/h for car passengers and $4.68/h for light truck passengers for ridesharing to be an attractive alternative in the average case. Jacobson and King also state that ridesharing can also be made more attractive by increasing parking fees and road toll costs. If parking fees and road toll costs of $1 are added to each vehicle trip, the maximum rational value of time for travelers to choose ridesharing approximately doubles, to a value of $9.05/h for car passengers and $8.68/h for light truck passengers. More substantial increases in parking fees and road toll costs can make ridesharing the most rational economic choice for many travelers.

Therefore, an effective promotion of carpooling practices must address all potential benefits (e.g., share of costs, lower road tolls, lower parking fees, etc.) and drawbacks (e.g., loss of time for pickups, lack of reliability in other users, lack of a large bourse of users for destinations with limited or nonexistent public transport, etc.). Experiences conducted so far also show that an active involvement of a third-party organization (e.g., state administration, a large private company with a bourse of commuters, etc.) is desirable. The 3- or 4-Helix model is therefore critical.

5 The FBC Business Model: Building a Critical Mass

The key challenge, with reference to the use of carpooling systems and web 2.0 tools in general, is attainment of a critical mass of users. To this end, FBC has invested significant effort in building the requisite start-up relationships in different parts of the system. With reference to the local market (Girona), agreements are in place with the University of Girona and the *Ajuntament de Girona* (Girona City Hall) with discussions ongoing with other strategic level partners necessary to build the necessary infrastructure for the network. This Girona "start-up" case is detailed later.

One possible strategic avenue to building critical mass (see Fig. 3.5 below) is to begin with the student community, more open to sharing car journeys with people they do not know, and also more likely to experience financial pressures related to mobility. As such, students are viewed as the perfect early adopter for carpooling, and already form the basis of several competing firms' strategies such as Zimride[7] in

[7] http://www.zimride.com.

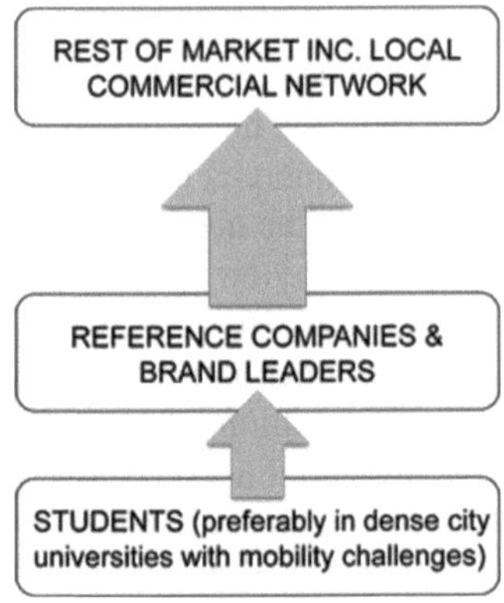

Fig. 3.5 Diffusion strategy

the USA. The aim is to build a type of "cult" following within the student population (for us in the short-term, this means the Catalan and then the Spanish student population), which links well with low-cost viral marketing possibilities through Facebook.com and other social webs. A large percentage of these students also become the professional classes of tomorrow, which mean that large student communities attract the attention of reference brands, attracted by a localized critical mass and future market. In turn, these brands pull through the rest of the commercial network, sustained in the longer term by the dual social currency system of FBC.

Based on a stakeholder analysis, the following key partner types for each potential FBC location are identified, focusing on the Catalan and Spanish markets.

Technology partners (strategic private or investor stakeholder): Related to GPS/mobile phone consumption, and perhaps also R&D of the system. Potentials include Telefónica, Vodafone, Ericsson, and other smaller providers such as Lleida.net and FonYou.

Industry partners (strategic private stakeholder): Those with a strong reputation in the industry and the potential to help FBC attain critical mass through the access to their present customers. Potentials include those from the automotive and insurance industries such as SEAT, and RACC, and smaller providers such as Rodi.es and Petrocat.cat.

Social partners (community customer stakeholder): Local "hotspots". These may also be large enterprises, which could help build a critical mass, but who may not the system for their own employees, such as supermarket Caprabo, or other key centers such as Universities.

Commercial partners (sponsors stakeholder): Related to a network surrounding the use of complementary currencies (NuRide). Potentials include all commercial establishments in Barcelona, of special interest being those who are attractive to the Hot Spot population, and/or who also have establishments in other territories for future potential expansion of e-Hitchhiking.

Marketing partners: Considering that the creation of the requisite number of commercial agreements would be a difficult task for a small start-up company, this considers assistance in the building of a strong commercial network as described above.

Policy partners (strategic public stakeholder): Public administrations that are in the position of promoting the use of the service through local infrastructures, such as the Barcelona City Hall and Chamber of Commerce.

5.1 Implementation and Roll-Out

The FBC domestic market entry begins with the markets it knows best which is critical in building solid partnerships that will sustain the network and brand. This means Girona first, and then Barcelona. After both locations are running effectively the vision is to expand to other reference cities in Spain, namely, Madrid and Valencia, followed by one small and one large European city in 2015 and three large international cities in 2016. The main steps toward implementing FBC are as follows:

- Identification of a suitable city.
- Identification of key hot spots.
- Identification of key partners according to typology.
- Engagement with key partners by the internal team.
- Engagement with commercial partners by external agency.
- Live implementation and connection to users (viral campaign).

Viral marketing is one of the most efficient promotion tools in terms of costs and capacity to reach a wide range of people. Individual potential users are the people who actively spread the message, talk about the product and arouse interest in the community, creating a buzz around the product. This requires the introduction of an attractive message, communication about the product (or at least arousing curiosity) through channels that should easily spread the message among the people who represent the potential users. With web 2.0, it is easy to generate such a buzz, if the message is attractive and well spread. There are several specialized webs, blogs, social networks, portals, and forums on almost every topic, and the objective is to choose the right place to put a message for a specific product.

One of the main advantages of the current FBC project is that its scope is multidisciplinary: on one side, it is a tool that uses the latest technologies including social networks and mobile devices. On the other hand, it is linked to sustainability, the environment and transportation; in total representing several large user communities. In addition to traditional ways of spreading a message and promoting a product via Twitter.com, Facebook.com, newsletters, or RSS feeds, additional channels must be found to stand out from competing messages. Currently, there is a significant amount of websites and blogs focused on the two major groups of technology and environment, so these channels are a good way to reach FBC users.

5.2 *Vision in Action: Start-up Case: Girona, Catalonia, EU*

FBC is currently building a strong commercial case for successful implementation in Girona during the third quarter of 2011, supported by the necessary strategic partnerships. Building on the e-Hitchhiking platform, the new version (http://www.mobilitatsostenible.cat) begins with a potential user population of over 15,000.

- *Social partner*: The University of Girona viewed the original e-Hitchhiking application of FBC as a strategic project within their upcoming implementation plan on the theme of University Campus of Excellence in Spain. They will continue to support the full implementation case through building a network of commercial enterprises and dissemination.
- *Technological partner*: Lleida.net will be the text message provider used to facilitate communication between passengers and drivers.
- *Industrial partners*: Advanced discussions have taken place with the car service and repair center Rodi.es as an industry partner for FBC to offer special discounts on their services, and with petrol company Petrocat.cat. Rodi has operations centered on Catalonia but also with a growing presence in other areas of Spain. Petrocat is focused on the Catalan market through a chain of petrol stations and 20 distribution centers, with the added stability also of being part-owned by Repsol. The advantage of having a small to mid-size company in this area is the interest they have in supporting the implementation of FBC in new territories where they are also trying to establish a foothold.

6 Final Discussion and Conclusions

Collaboration and networking for innovation is more than summing efforts. It is about creating synergies among and across participants. Resulting partnerships should be robust and long-term oriented, economically and socially sustainable, and, at the end of the day, human. This chapter tries to show an example with FBC.

In the present day, the complex world is characterized by high uncertainty and difficulties in predicting future strategies, and traditional business models are often replaced or complemented with innovative solutions. Information Technologies (ICT) tend to offer a variety of solutions and electronic, and "e"-versions of traditional products/services emerge.

Recent trends in the field of innovation target "soft" aspects compared to traditional product/process innovations. Organizational innovation, service innovation, eco/green innovation are, to mention just a few, targeted toward satisfying non-purely financial performance. It is in this scenario where social innovation gains importance, and companies start to exploit this area. Moreover, coworking, collaboration, and networking also seem to be interesting options in an environment characterized by resource and knowledge scarcity.

Our aim in this chapter is to illustrate an initiative combining all these aspects—ICT, networking, innovation, societal and environmental benefits—and insist on its benefit to society in general and participant stakeholders in particular. The 4H approach is appropriate and useful in the sense that it covers the main actors and the linkages established among them, applied to the "e"-version of traditional carpooling. This chapter shows how FBC is deployed for supporting the carpooling communities, articulating the help of public institutions, sponsors, businesses and individuals. It contains social, economic, administrative and technical visions.

The focus of FBC is on sustainability in terms of long-term value creation, on competitiveness, and how different entities (business, academia, and government) achieve this and add value through stronger networks and connectivity that better involve and engage different actors—companies, customers, and other knowledge generators. FBC could therefore be viewed as putting the 4H into practice, as it seeks to engage public administrations, users, private enterprise, and others around a common cause, which also has the added bonus of being socially responsible.

One of the main advantages of the technology behind FBC is that its scope is multidisciplinary: on one side it is a service that uses the latest technologies including social networks and mobile devices. On the other hand, it is linked to sustainability, the environment, and transportation, in total representing several large user communities, spreading and promoting a positive message through social networks.

Acknowledgments This research is funded by the *grup de recerca consolidat* CSI-ref.2009SGR-1202. We also wish to acknowledge the work of Dr. Steven MacGregor in the strategic development of FBC and his insight into its fit with corporate social responsibility.

References

Afonso, O., Monteiro, S., Thompson, M. (2010) A Growth Model for the Quadruple Helix Innovation Theory, Working Paper Series, NIPE WP 12/2010.

Arnkil, N., Järvensivu, A., Koski, P., Piirainen, T. (2010) Exploring Quadruple Helix – Outlining user-oriented innovation models. Työraportteja 85/2010 Working Papers.

Jérôme Blanc (2011), Classifying "CCs": Community, complementary and local currencies' types and generations, International Journal of Community Currency Research 15:4–10.

Carayannis, E., Campbell, D. (2009) 'Mode 3' and 'Quadruple Helix': Towards a 21st century fractal innovation ecosystem. International Journal of Technology Management (46) Nos. ¾: 201–234.

Commission of the European Communities (2003) *The role of the universities in the Europe of knowledge*. Communication From The Commission, Brussels, 05.02.2003, COM (2003) 58 final.

Commission of the European Communities (2007) *GREEN PAPER Towards a new culture for urban mobility*, Brussels, 25.9.2007 COM (2007) 551 final.

European University Association (2011) *Response to the communication from the commission – The role of the universities in the europe of knowledge*. http://www.eua.be/typo3conf/ext/bzb_securelink/pushFile.php?cuid=2122&file=fileadmin/user_upload/files/EUA1_documents/EUA_Role_Univ_Response.pdf. Accessed on May 2011.

Ferguson, E (997). The rise and fall of the American carpool: 1970–1990. *Transportation*, 1997, 24(4):349–376.

S. H. Jacobson, D.M. King (2009), Fuel saving and ridesharing in the US: Motivations, limitations and opportunities, 2009, *Transportation Research Part D* 14: 14–21.
Gregory A. Krohn, Alan M. Snyder (2008), An Economic Analysis of Contemporary Local Currencies in the United States, International Journal of Community Currency Research (12):53–68.
Levin, I.P. (1982) Measuring trade offs in carpool driving arrangement preferences. *Transportation*, 1 1(1):71–85.
B. Lietaer (2001) The Future of Money: a new way to create wealth, work, and a wiser world. Random House Australia (Pty) Limited. 2001. ISBN 0 7 126 9991 0.
Lindqvist and Tegner (1998), in Gross et al., 2009, Lindqvist, E., & Tegner, G. Measures to reduce the CO_2 emissions from the transport sector in the city of Stockholm. In. *Proceedings of the 1998 4th international conference on urban transport and the environment* (Pages. 349–358). Ashurst, England, Lisbon, Portugal: Computational Mechanics Publications. http://dynamicridesharing.org.
G. Santosa, H. Behrendtb and A Teytelboymb (2010) Policy instruments for sustainable road transport – Part II, *Research in Transportation Economics*, Volume 28 (1): 46–91.
Seyfang, G. (2009). The New Economics of Sustainable Consumption. Basingstoke: Palgrave Macmillan.
Seyfang, G. & Smith, A. (2007) Grassroots innovations for sustainable development: Towards a new research and policy agenda. Environmental Politics, 16 (4): 584–603.
James Stodder (2009) Complementary credit networks and macroeconomic stability: Switzerland's Wirtschaftsring, Journal of Economic Behavior & Organization, Volume 72 (1): 79–95.

Chapter 4
Collaborative Healthcare Innovation in Sweden

Andreas Larsson, Susanna Bill, Jenny Ingridsdotter, and Annika Olsson

Abstract The medical technology (medtech) industry in Sweden is situated within a complex innovation ecosystem, in which various stakeholders from the public, private and academic sectors need to collaborate to meet demands on effective and efficient healthcare. Demographics are changing and those in need of healthcare are not only larger in numbers than ever but they are also more knowledgeable and demanding. Increasing innovative performance is crucial in both the private and public healthcare sectors, but bold steps forward need to be taken in light of stricter rules and regulations for how healthcare stakeholders should manage both their internal processes and the ways in which they interact with other stakeholders in the larger innovation system. The traditional way in which medtech companies gain access to user needs, primarily working through a sales–purchasing relationship with the public healthcare sector, is outdated and needs to be replaced with an increasingly collaborative and cocreative model of healthcare innovation.

This chapter describes experiences and lessons learned from InnoPlant, a 3-year (2008–2011) action learning project involving three companies from the Swedish medtech industry, two county/regional councils responsible for public healthcare, and four academic institutions—carried out within the framework of the Swedish Product Innovation Engineering program (PIEp). The purpose of the project was to advance the capability of stakeholders from the public, private, and academic sectors to collaborate in the cocreation of healthcare innovations.

A. Larsson (✉) • S. Bill • A. Olsson
Department of Design Sciences, Lund University, Box 118, Lund 221 00, Sweden
e-mail: andreas.larsson@design.lth.se; susanna.bill@design.lth.se; annika.olsson@plog.lth.se

J. Ingridsdotter
Baltic and East European Graduate School, Södertörn University, Huddinge 141 89, Sweden
e-mail: jenny.ingridsdotter@sh.se

S.P. MacGregor and T. Carleton (eds.), *Sustaining Innovation: Collaboration Models for a Complex World*, Innovation, Technology, and Knowledge Management,
DOI 10.1007/978-1-4614-2077-4_4,

1 Introduction

The medical technology (medtech) industry in Sweden is part of a complex innovation ecosystem, in which stakeholders from the public, private and academic sectors need to collaborate closely to meet increasingly higher demands on effective and efficient healthcare. People in need of healthcare not only come in greater numbers than ever but they are also increasingly knowledgeable about the healthcare system and are demanding healthcare that is truly attentive to their particular needs. To meet higher demands, it is crucial to increase the innovative performance of stakeholders in the private and public healthcare sectors, but advances need to be made while adhering to stricter rules and regulations for how stakeholders should manage both their internal processes and the ways in which they interact with other actors in the innovation system (Herzlinger 2006). For instance, the *Swedish Public Procurement Act* (Swedish Competition Authority 2011), put in place in 2007, sets a rather strict framework for the interaction between healthcare stakeholders during procurement. During the procurement negotiation phase, either party (i.e., the healthcare organization and the medtech company) is only allowed to make contacts with the other party for clarification purposes. This means that the medtech company may not contact any other personnel than the appointed responsible person at the healthcare purchasing organization. This newly imposed framework needs to be viewed in a historical light, since the Swedish healthcare system has a long tradition of collaboration between medtech companies and the public healthcare sector, with new products and solutions continuously created as incremental innovations based upon clinical insights. Taking the Public Procurement Act into consideration, the medtech companies need to master a balancing act concerning how to successfully create and maintain strong relationships with the public healthcare sector to access relevant user insights from healthcare practice without breaching the requirements of the Public Procurement Act.

In addition, the concept of "user insight" is somewhat difficult to define precisely in the public healthcare sector, since there are many stakeholders who might be viewed as users depending on the context. In the Swedish regional/county councils, which govern the hospitals and primary care centers, there is a need to improve the understanding between the purchasing and usage structures related to the design, development, and procurement of new healthcare devices. The purchasing structure is largely characterized by a business-to-business logic, where decisions are primarily related to how well medtech suppliers can meet demands in aspects such as technical performance, reliability, and cost. However, user needs are more multidimensional than that, and they include the needs and requirements of a wide range of healthcare practitioners and administrators that are dependent on new equipment to deliver both effective and efficient healthcare. Too often, a lack of understanding between the purchase organization and the use organization results in equipment that poorly fulfills usage needs, although the equipment might actually has been delivered in accordance with the agreed requirement specifications. The way in which the medtech companies gain access to user needs, within the boundaries of

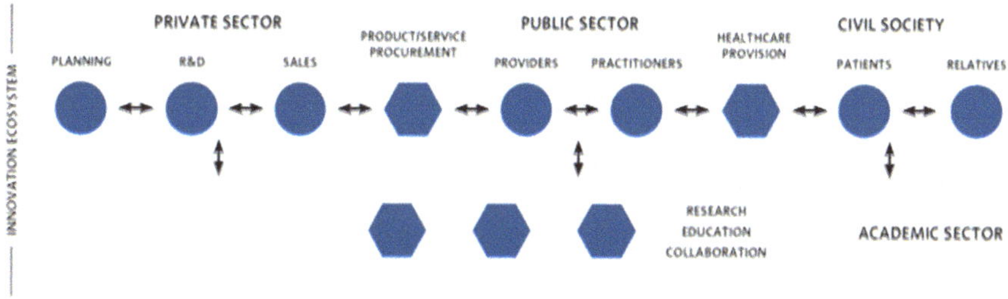

Fig. 4.1 The healthcare innovation ecosystem

the Public Procurement Act, is primarily through a sales–purchasing relationship, meaning that important user insights may never reach the new product development functions of the medtech industry companies.

The healthcare innovation system demands that existing technology is quickly adapted to new circumstances, and that entirely new technologies are developed to meet new and higher demands. Since the development lead-time can be relatively long—an incremental change can take as much as 2 years to bring to market—it is important to deeply understand and quickly act on user insights for both incremental and radical innovation purposes. Involving a wide range of users in the front-end innovation work allows user needs to influence the development work from the start, thus minimizing the risk of a mismatch between what new medtech devices offer versus what users actually need. In the context of the larger healthcare innovation system, involving users actually implies establishing a closer relationship with a range of stakeholders for a range of purposes.

Figure 4.1 shows a schematic overview of the various stakeholders in the overall Swedish healthcare innovation system. In different ways, each stakeholder contributes to the shaping of innovations. Note that this is not a complete picture. For example, there are supply chains in both the private and the public sector that are implicit here. Also, communication flows within the innovation system are more complex than the arrows show. However, the figure aims to highlight that there are several boundaries to cross to achieve collaborative healthcare innovations. For example, public healthcare is a business where customers and users are seldom the same. How can we become better at cocreating, selling, and buying innovations that fulfill different stakeholder needs in effective and efficient ways? Is it possible to provide better healthcare at a lower cost? Further, one could argue that public healthcare is a business where patients tend to return despite poor past experiences. How can we become better at treating our citizens more as valued customers? In Fig. 4.1, patients and relatives are included under "civil society" to highlight that the medtech devices and healthcare services ultimately need to provide value to those that are in need of healthcare, and their families.

This chapter describes experiences from InnoPlant, a 3-year (2008–2011) action-learning project involving three companies from the Swedish medical technology industry, two county/regional councils responsible for public healthcare, and four academic institutions—carried out within the framework of the Swedish Product

Innovation Engineering program (PIEp). The purpose of the InnoPlant project was to advance the capability of stakeholders from the public, private, and academic sectors to cross boundaries and collaborate in the cocreation of healthcare innovations. The chapter adds to the results from earlier studies by Olsson et al. (2010) and Bill et al. (2011).

2 Method

The main idea of the InnoPlant research project was that each of the two county/regional councils and three medtech companies would designate one innovation project, which would be supported by academia in an action research-inspired process. The experiences and learning outcomes from these projects would then be brought by representatives of each organization into "learning network meetings," where these experiences and lessons learned would be shared, questioned, further explored, discussed, and reflected upon together with the other organizations.

The action learning and action research methods used in this research project have combined several approaches for the common purpose of inspiring participating organizations to reflect on current ways of working and to increase their innovation capability through trustful, learning partnerships with other stakeholders in the healthcare innovation ecosystem. Implementing sustainable new ways of working in the participating organizations puts high demands on the organizations and the individuals acting in the project, and it demands strong support from the upper management sponsors in the respective organizations.

The public healthcare sector has been represented by two county/regional councils, the medtech industry by three companies, and academia by engineering faculties from two universities and a faculty of social science from a third university. A fourth organization has also been involved in the project, a joint research center formed by two universities and one county council to promote research and commercialization activities in the field of medical technology. All parties are located in Sweden, and their roles can be described as follows:

- *Engineering faculties of universities*: Three professors, four senior researchers, and three PhD students from product design and development related departments represented the engineering faculties of two universities located in different parts of Sweden. The role of academia in the learning network has been to facilitate meetings between different stakeholders in the project (and in the innovation system) and to turn questions for reflection into new actions. Researchers have taken part in the action research projects in each organization with the intent to increase the partners' innovation capability. The researchers' role in the learning network has been to facilitate a learning situation by raising relevant questions and perspectives.
- *Social science faculty*: A Master's student tutored by a professor from an ethnology research group represents the social science faculty at one university. The role of the ethnologist has been to perform a meta-level study on the research

project to describe how the culture within this type of project plays out and can be further developed. Furthermore, the ethnologist has continuously provided feedback to the other participants in the research project regarding, for instance, working climate and working procedures.

- *Medtech industry*: Three medtech companies have been involved in the project, and they are developing and commercializing products within the fields of mobility devices, sterilization equipment, and anesthesia systems. As noted earlier, each company has brought one innovation project into InnoPlant. In the learning network, there were two representatives from each company, holding positions of research and development manager or product manager in their organization.
- *Public healthcare system*: The two county/regional councils have provided one product innovation project each, related to the development of a tool for heart failure diagnosis and an IT system in healthcare. A project was also initiated where certain inventive employees within the public healthcare system were invited to participate in workshops considering the conditions for realizing product innovations in public healthcare. In the learning network, each county/regional council was represented by a practitioner working with the innovation project and someone who works with strategic issues regarding innovation.
- *Joint medtech research center*: The assistant director of the center was the project coordinator for InnoPlant and was thus responsible for the planning and coordination of the learning network meetings.
- *Guest lecturers*: Different guest lecturers were invited to introduce a selected topic for each meeting in the learning network. The lecturer was either a practitioner with relevant experience or a researcher with interesting research results within the topic of collaborative healthcare innovation.
- *Steering committee of directors*: The steering committee of directors included professors, CEOs, and directors from the participating organizations. The role of the steering committee was to monitor the project progress and provide top-down support for the actions taken in the organizations, as well as disseminating and communicating the results within their own organizations.

As can be seen in Fig. 4.2, the above actors and roles were combined in three main approaches to collaboration in InnoPlant. The three sectors were represented at each level of interaction, but the degree of participation was different. It should be noted that this figure is a conceptual sketch of the collaboration approaches. For instance, although the action learning network meetings provided occasions in which all partners could participate in a collaborative spirit, some of the more fruitful outcomes were found in the dynamics between the local research and innovation projects (where researchers and practitioners worked closely together on more specific issues than in the more generic network meetings).

1. *Action learning network*: Learning network meetings involving all parties were organized three times a year. These meetings normally started in the evening with a guest lecture, and continued over the next day with workshops. These network meetings had two primary objectives. The first was to feed in knowledge from recent research projects in a field that the participant organizations

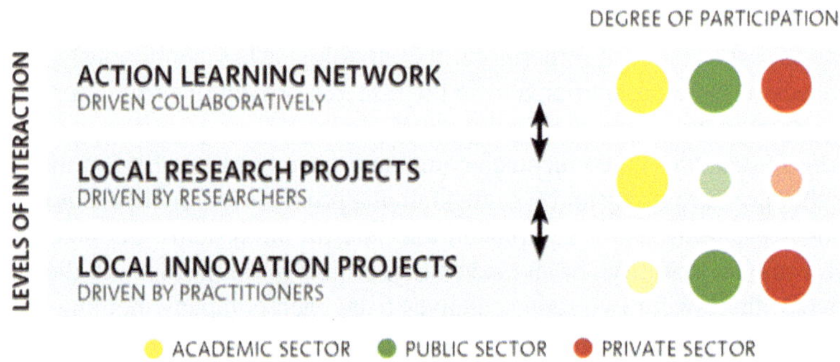

Fig. 4.2 Three approaches to collaboration in InnoPlant

had identified as particularly interesting. The researchers could disseminate knowledge about research in their own fields of interest, and they could also invite experts if the chosen topic was out of their own expertise area. The second objective was to report the experiences from and set new goals for the local projects, including feedback and benchmarking activities with colleagues from the other public and private sector partners. This second part included the sharing of learning outcomes between and within the participating organizations and between and within the academic partners. The researchers' role was to facilitate and turn questions for reflection into new actions.

2. *Local research projects*: Depending on the current needs of the organizations, local empirical studies were used to trigger change processes and to coach the auditing of innovation capabilities in each organization. The academic researchers drove these studies, in close collaboration with the local partner representatives from the medtech companies and county/regional councils. The studies contributed to broadening the commitment and learning in the organization, which created opportunities for doing things differently on a local level.
3. *Local innovation projects*: These projects were chosen by each organization as needed, and the aim and scope differed between product innovation projects, organizational development projects, or other types of business development projects.

The network organization chosen in InnoPlant can be seen as having two primary aims. The first one is to facilitate innovation capability within the participating organizations, and the second one is to develop sustainable relations and trust between researchers from participating universities and participating organizations from the public and private healthcare sectors.

Throughout the InnoPlant project, both physical meetings and telephone meetings between participants were monitored by an ethnologist to better understand how the cross-cultural communication played out in the project's day-to-day activities and to explore how to better facilitate this boundary-crossing communication in both ongoing and future projects. Departing from the notion that the three sectors (i.e., public, private, academic) taking part in the project have their own social codes, different approaches to the concept of innovation, and their own cultural solutions to problems, the ethnologist (one of the coauthors) studied which cultural patterns can be traced in the participants' behavior and the ways they relate to each other and to the project. In addition to a detailed observation of the encounters between participants, qualitative interviews were also conducted with the participants, which concern their reflections and opinions about the overall project and their own local projects. These interviews provide the empirical base of the excerpts provided later in this chapter.

3 Learning Networks

Learning networks can be defined as a network formally set up for the primary purpose of increasing knowledge (Bessant and Tsekouras 2001). Actors in such networks provide an arena for experience exchange and learning where "old truths" can be challenged and new perspectives formed (Bergh 2009). According to Bessant et al. (2003), learning networks may encompass different learning targets, such as: increased professional knowledge and skills, improved awareness of a particular field, improved knowledge regarding regional interests, and sharing knowledge on how to do a particularly novel task. Learning networks can also be seen as an arena for the use of different learning methods, and they provide a good basis for inter-organizational learning to take place. A learning network setup provides opportunities to exchange experiences on new theories and existing models relating to the issues that the participating companies are working with (Ritzén et al. 2005). In a learning network involving academic researchers, the participating organizations are responsible for initiating change, while the researchers facilitate the process of initiation, dialogue, participation, and reflection (Rasmussen 2004).

The interaction between academics from different disciplines and practitioners from different settings in learning networks creates a learning context for all parties, which supports learning and creation of new knowledge that could be generally applicable for the involved parties. Engaging in this type of learning process could therefore be useful and enriching for the learning, professional development, and competence of academics (Karlsson et al. 2007).

4 Experiences and Lessons Learned

One of the most important experiences drawn from the initial phases of this action learning project is that progress in a project that involves more than one organization requires a comprehensive trust-building process between the participating organizations and the individuals involved (Olsson et al. 2010). This trust building was evident during the whole of the project with a clear evolution in the levels of trust from commitment to companion, and finally, to competence trust (Bergh et al. 2011).

The commitment trust-building was accomplished through the kick-off meeting, where the different participants had the opportunity to get to know each other on a personal level, and the setup was planned to enable the participants to perform trust-building and team-building activities. The first meeting was set up at a neutral location outside of the participating organizations (i.e., not at the location of any participating party), to create a physical space that was new and unknown for all and thereby not considered as anyone's territory. The aim was to develop both commitment and companion trust, although the level reached at this first meeting was only at the commitment level. In the second meeting, hosted by one of the industrial partners, the active trust-building process started on the companion level, and the atmosphere "opened up." Potential reasons for this were that the participants knew each other after the first meeting, and that the hosting organization took an open-minded approach, presenting both strengths and weaknesses in their way of working with innovation. The organizations hosting the learning network meetings stimulated the trust-building process and engagement by being open and thus inviting others to be equally open about their innovation capability (or lack of it).

Seen from an ethnologist's perspective, collaboration is never easy, and cross-cultural communication is often accompanied by misunderstandings. To say that there are three different cultures participating (i.e., public/private/academic sectors) might actually be too simplistic on account of the many differences and nuances within them. For instance, the researchers have different academic backgrounds and disciplinary commitments, the participants from the public sector represent different regions and hospitals, and the participants from the private sector correspond to different business cultures. Therefore, it is impossible to speak in general terms about these three groups. Nevertheless, to enhance understanding between the different sectors in projects of this nature, the ethnologist in our team has tried to simplify complex cultural aspects into general terms, speaking of academia, medtech industry, and public healthcare as rather homogeneous cultural units.

The ethnographic study was ultimately intended to facilitate the process of collaboration, and material was gathered to assist the search for crucial communication and collaboration challenges within the InnoPlant project. Those challenges that were found to be particularly significant have been used for this chapter. The aspects chosen are mainly those who might interest groups facing similar tasks. The ethnographic material collected provides an opportunity to understand how the InnoPlant project has developed over time and how participants have experienced it. The focus of the ethnographic research has been to seek out challenges to improve communication and collaboration. Although the other research methods have shown that the project has also resulted in a number of best practices and success factors to guide triple helix collaboration, this chapter reports primarily on the challenges and lessons learned uncovered in the ethnographic study. We made this choice because we want to underline that collaboration is an ongoing learning process, and we want to provide an honest report on some of the stumbling blocks along the path to successful collaboration, rather than merely showing the end result.

In order to ensure anonymity, we have constructed quotes merging various statements of participants from the same "culture." We investigate the question: what do they think about central cultural and communicative aspects of this collaborative project and how should projects like this be performed?

4.1 Conditions and Expectations

A major complication has been the fact that the conditions and expectations of the three cultures differ from one another when entering a project such as InnoPlant, and therefore, the absence of a joint goal has led to frustration during the process. A more profound discussion of every organization's presumptions, practices, and expectations has been lacking, resulting in a continuous search for meaning, which is related to the confusion concerning goals and purposes. Anyone who has taken part in boundary-crossing projects would probably understand that there is an inherent "culture clash" that is truly challenging.

InnoPlant is an innovative project planned around concepts of action learning and the idea of generating new knowledge. Consequently, to communicate the overall project vision and to work out a common definition of what "new knowledge" means, or an explicit notion of the different results expected by every organization, has been one of the key challenges. As many participants have pointed out in interviews, the main interests of each group diverge—companies seek economic profit, researchers create knowledge, and the public healthcare sector administrate healthcare. The difficulty of InnoPlant, and in most cross-boundary collaboration projects, resides in the fulfillment of all partners' desires. How could common satisfaction be achieved? Here are the voices of some participants:

> Academia: This is a project of change involving learning, it has to take time, and we have to make room for social aspects. It is important to create a time and a space beyond the ordinary everyday practice. It is hard for us not to act as consultants. We also have to consider that it's important for us to publish results and create knowledge.

> Public healthcare: The purpose should be explicitly outspoken. It has been really hard to understand what we are actually doing and where we are going. Maybe it's a conscious strategy that it should be unclear so that we look everywhere—the insecurity of the group implies an increased split vision, but it's been a bit too fuzzy. If that's the purpose, it should be explicit. Now we just basically feel lost.

> Industry: From the beginning, the purpose was very ambitious. It is hard to see what we can change and how we should measure it. It becomes more of research study, but maybe that's fine, as long as it is about constructing knowledge maybe it isn't that important to get explicit results. It was hard for us in the beginning, but maybe this is enough.

> Industry: Is the purpose really that we should drink coffee and have a pleasant time? I don't know how to get a return on the investment. It's a bit unclear where we're heading to in the project. There's an ethical difference here that should be considered as well and it has to do with the fact that we are actually here to make money and the public healthcare to save lives. It's hard to see what this is bringing back to us right now. Instead of accomplish something and see if it worked out, as in standard consulting manner, in this project we have to analyze the analysis and that makes it hard to create value in the short-term. That's the challenge.

As is visible from these merged statements, the different organizations have very different purposes, conditions, and expectations when entering the project. To make it work, at least from an industry perspective, the goals should be well defined and the project management clear. This project has been planned by academia, and this means, for instance, that there is a tendency to value emerging research questions and a long-term perspective. What would have been different if the medtech or public healthcare partners had been responsible for the planning and management?

4.2 *Responsibilities, Meeting Cultures, and Cooperation*

In an innovative project such as InnoPlant, the focus on finding new ways of working has also meant that it has not always been clear what role the different participants should play. InnoPlant was designed with a vision of active participants; the researchers were expected to facilitate meetings and provide the environment for the participants to assemble. In other words, academia took responsibility for planning and facilitating meetings. This led the other participating organizations to feel that researchers where being passive and not interacting in discussion as much as they would have liked them to. At the same time, researchers felt that the other organizations were not assuming responsibility for carrying out activities. This issue was discussed at the fourth network meeting and led to an initiative to plan subsequent meetings collaboratively by researchers and representatives from one or more of the other partner organizations.

However, the three different cultures also have different routines and meeting practices, even if the events were intended to be jointly planned. For instance, some activities were experienced as highly abstract by members from the industry, while members from the academia commented on the same activities as very concrete and practical. Another issue has been how to let everyone be heard and not interrupt other speakers. A further aspect of cross-cultural communication in meetings was the use of central concepts such as "user" or "innovation," which are defined differently by each culture, and even within the cultures. As a researcher put it: "some of the problems are about translations, there should be more work done on the practical implications of the concepts." Regardless of interpretations of words, one of the greatest challenges concerned how to distribute responsibilities and how to collaborate across these boundaries:

> Academia: All responsibility was laid upon us, it would have been good if it was shared more, and if we had had a professional project management with the possibility to delegate tasks to industry.

> Academia: Sometimes it's hard to let go of the reins and allow everyone in. The industry does not have time to take responsibility. Sometimes the meetings have been really slow, and it's been too easy to hide behind Power Point presentations.

> Academia: Usually the industry has quite high expectations when they enter a project with a vague question formulation like this, and then they think that the researchers will arrange everything and tell them what it is all about; afterwards they notice that the researchers

don't know either but they have ideas about how to solve this together with new methods. Then the industry realizes that they have to work a lot, but they do not really have time or energy, and at the end they haven't done as much as they thought they would.

Public healthcare: There has been little discussion on concrete things and much on theoretical things, sometimes a bit too abstract. The project has been marked by a silent conflict as the academia pulled back wanting to study the process rather than taking part in the cooperation. There was also a conflict at the beginning as the industry regarded us as customers.

Public healthcare: A project like this needs more people from healthcare; the industry talks a lot and occupies a lot of space. Since the academia was arranging everything from the beginning, they assumed the role of the hostess, so the rest relaxed instead of taking responsibility for the project.

Public healthcare: Our local project is working out well, but I am surprised to see that the overall project is still a bit unstable, and the academia has such a hard time stepping into the arena as actors, they are always pulling back from participating, and I don't understand why. Otherwise it's an open culture with a lot of curiosity, and I think there have been possibilities to step up and speak of what's on ones mind.

Industry: It's been hard for us to take on a lead role in this project. Project management and methods should come from the academia. There has been an unclear distribution of responsibilities, and no one seems willing to take on project management, if the academia just tells us what they want we can deliver.

Industry: The academia has a hard time staying in the present; they don't stop to focus on a problem, but are always planning the next event, suggesting something new or having some kind of spin off.

4.3 Approaches to Time

Another central cultural aspect involving collaboration is the participants' approaches to time. Every culture has their own time concepts and values, which might be a challenge to the collaboration between the organizations. Here are some of the participants' opinions about the length of the project:

Academia: Three years is sufficient for a project like this; it's a normal time for a research project. You can't do much in shorter time. Maybe it should be a couple of years more so that there was more time to test other things.

Academia: Everyone is really enthusiastic during the hours we spend at the network encounters, but then when we are not there, we are just as busy as usual. We all have really full agendas, and it takes time to keep up in a project like this, not to speak of writing about it.

Public healthcare: We have a more protected environment, and we don't suffer the same time pressure as industry does. But at the same time, we can feel that it is a little too long and too quiet between the encounters. There is a cultural difference here; we want more action, and the academia wants to turn every rock over to make the project last the whole time. There are different views on action and reflection within the project.

Industry: It's hard for us to work on such long terms; 3 years is a long time, and this is really a challenge to InnoPlant since time gets so fragmented as you work in parts, and therefore it gets secondary. And when such a long project is not clearly defined, it gets even harder. We understand that the researchers want to study the process, but in our world, it gets difficult; innovation is a vague concept, and it's easy to lose focus in such a long-term project.

Another frequently recurring issue in the interviews is the lack of time. No one participates in the project on a full-time basis. It was planned with the presumption that everyone would engage emotionally in a way that was not realistic within the given time limits:

Public healthcare: The main project has suffered from a lack of continuity, and it's been quiet between the encounters. Our local projects are doing fine, but it seems as if the main one isn't really self-driven. Our organization has a more relaxed and permitting approach to time and its limits.

Industry: I don't have time for the missionary work a project like this really needs. If you really collaborate, you have to engage a lot of time. This project has served as a way to scratch the surface. Successful projects are a contribution to the company, but if the aim is only to study—and that means to drain the company of knowledge—it is difficult for us to provide time.

4.4 Lessons Learned

The focus of the ethnographic account has been challenges to a project of collaborative nature. Since we also wanted to get participants' opinions on what could be done differently in the future, we asked them to tell us about what they learned in the project and what they think about the future of similar projects:

Academia: Sometimes there is too much "us" and "them." In order to succeed, everyone has to see that it is a joint project, and that we should cooperate in planning. It's been really interesting to be at every organizations site and see how they work. The times that we've been around looking are when we've had most energy in the group.

Academia: It has been really enjoyable to do research in a group, create new relations, and get to know people in the field. It is important that the representatives from the different organization are in positions where they can transmit what they have learned to their organizations.

Academia: If a project like this is to be successful, it has to be based on concrete planning; to strengthen the innovation climate is to start at a meta-level. It's important to discuss early what the core of the project should be. It would also be good if the industry could enter the project with its own project leader that could cooperate with that of academia.

Public healthcare: We would have liked to have more discussions in smaller groups. At first, I hesitated to enter this project, but now I can see that it's been an interesting journey, and that I've profited of it in a way I didn't expect. I can definitely see a continuation of the project but it will have to be in a different form. Of course, everything depends on the fundamental conditions in our organization.

Public healthcare: One of the things that the project has given me is the network and insight into this kind of question formulation. I feel I know a lot more about the industry, how they reason and about the concept of innovation. Now, when I go to different meetings, I run into people from InnoPlant, and I have a bigger understanding of these issues, and my competence has extended. It has been a personal process of learning for me.

Public healthcare: When we have been around looking at different locations, it's been really interesting, and you could see that people get ideas. I think a project like this has to be more concrete and take place on the floor of every organization. The encounters in InnoPlant have offered the conditions for our different organizations to get to know each other. Now we will see how we go about cooperating outside of this structure.

Industry: You have to define the problem early, use the right methodology, and not lose focus. It is better to start ten projects and deliver ten, than to start 75 and deliver zero. To see how they do things in other companies and industries gives another dimension to your daily work. The form of the network encounters has been a lesson in itself. Innovation is about creating networks that work, and we have really had a chance to practice that here.

Industry: I think that one should have very concrete projects in order to pull something like this off. I also think that it has to do a lot with daring to invest time and resources. Everyone has the intention of improving things, but it's hard to get people to work with something that they don't really understand, it's better to put things straight.

Industry: I believe the project has been really positive and definitely something that could continue, but there has to be an explicit goal, what methods are we to work with and a project leader to report to. There should be a briefing every 6 months that we could work against.

5 Conclusions and Discussion

This chapter describes experiences and lessons learned from a 3-year action learning project involving three companies from the Swedish medtech industry, two county/regional councils responsible for public healthcare, and four academic institutions. The purpose of the project was to advance the capability of stakeholders from the public, private, and academic sectors to collaborate in the cocreation of healthcare innovations. However, as has been shown throughout the ethnographic accounts provided in this chapter, collaboration is not easy, and cross-cultural communication is laden with misunderstandings. Researchers have different academic backgrounds and disciplinary commitments, participants from the public sector represent different regions and hospitals, and participants from the private sector correspond to different business cultures. As many participants have pointed out in the interviews, the main interests of each group diverge—companies seek economic profit, researchers create knowledge, and the public healthcare sector administrate healthcare. The difficulty of InnoPlant, and in most cross-boundary collaboration projects, resides in the fulfillment of all partners' desires. Some participants have felt that researchers were being passive partners, not interacting in discussion as much as they would have liked them to. Researchers, on the other hand, have felt that the other organizations were not assuming responsibility for carrying out activities. Another central cultural aspect involving collaboration is the participants' approaches to time. Every culture has their own time concepts and values, which might be a challenge to the collaboration between the organizations.

Again, it should be noted that this chapter has focused on providing an honest account on some of the stumbling blocks along the path to successful triple helix collaboration, rather than providing an after-the-fact roadmap to perfect relationships between the academic, public and private sectors. Therefore, we have primarily focused on the overarching level of interaction (see Fig. 4.2), i.e., the action learning network level. We have seen that there are clearly different learning cycles and outcomes depending on which level of interaction we base the analysis on. It is quite clear to us that many of the challenges we attend to in this chapter are related to the action learning network level. At the levels of the local research and innovation projects, it seems like the various actors (usually from two out of three sectors) have usually been able to come to a shared understanding of objectives, activities and deliverables, whereas the aggregated level represented by the network meetings seems to have added a layer of confusion about aspects such as common objectives and expected outcomes.

Moving into a closure phase where we are planning how to continue the project beyond these 3 years, all partners agree that the learning and trust-building experience provided mainly through the network meetings is what has actually enabled us to create a strong network with the willingness, trust, and competencies to take the next step in triple helix innovation. This chapter shows that such a learning experience can often be hard-earned, and we have learned that our attempts at bridging different cultures are definitely worth the extra effort, much because it helps partners from various sectors to leave their own comfort zones in the collective search of new pathways to innovation.

Finally, the aim of this chapter has not been to make judgments about "right" or "wrong" approaches and perspectives, but rather to openly share experiences and lessons learned from a variety of perspectives in the healthcare innovation system, based on the belief that sharing expectations and continuously reflecting on the roles and responsibilities of each partner would improve the collective innovation capability of all partners in a triple helix system.

References

Bergh P (2009) Learning among entrepreneurs in formal networks: Outcome, processes and beyond. Doctoral dissertation, Umeå University.

Bergh P, Thorgren S, Wincent J (2011) Entrepreneurs learning together: The importance of building trust for learning and exploiting business opportunities. International Entrepreneurship and Management Journal 7, no. 1: 17–37.

Bessant J, Kaplinsky R, Morris M (2003) Developing capability through learning networks. International Journal of Technology Management and Sustainable Development 2, no. 1: 19–38.

Bessant J, Tsekouras G (2001) Developing learning networks. AI & Society 15: 82–98.

Bill S, van Helleputte J, Häglund F, Larsson A, Olsson A, Sjölund F (2011) Rethinking User Involvement in the Front End of Healthcare Innovation. Proceedings of the 18th International Product Development Management Conference (IPDMC'11), June 5–7, Delft, The Netherlands.

Herzlinger RE (2006) Why Innovation in Health Care Is So Hard. Harvard Business Review 5, 1–10.

Karlsson, J, Booth S, Odenrick P (2007) Academics strategies and obstacles in achieving collaboration between universities and SMEs. Tertiary Education and Management 13, no. 3: 187–201.

Olsson A, Wadell C, Odenrick P, Norell Bergendahl M (2010) An action learning method for increased innovation capability in organisations. Action Learning: Research and Practice, 7: 2, 167–179.

Rasmussen L (2004) Action research: Scandinavian experiences. AI & Society 18: 21–43.

Ritzén S. Zika-Viktorsson A, Forslin J, Engwall M (2005) Learning to change: A network approach to engineering management development. Proceedings of the International Conference on Engineering Design (ICED'05), August 15–18, Melbourne, Australia.

Swedish Competition Authority (2011) The Swedish Public Procurement Act – an introduction. Accessed 7 June 2011. http://www.kkv.se/upload/Filer/ENG/Publications/Public_procurement_intro.pdf.

Chapter 5
Understanding Collaboration in Knowledge Processes in Indian Industry

Gokula Vijaykumar A.V. and Amaresh Chakrabarti

Abstract Collaborative design supports quality innovation in reduced cost and time to market; this is critical to sustain organizations within the current competitive product development landscape. Understanding the knowledge processes that occur through collaboration among stakeholders in designing should help industry assess the quality of its collaboration and knowledge processes. Existing models for understanding knowledge processes during collaboration are inadequate in describing significant details of these processes; importantly, they do not stress the centrality of interactions in processing knowledge. A collaborative model called Knowledge-Requirements-Interactions-Tasks, or "KRIT," is proposed to help understand how collaborative knowledge processing takes place through interaction among stakeholders in product development. Also, an Influence model has been proposed to assess the levels of satisfaction of the four elements in the KRIT model. Indicators for satisfaction of knowledge, requirements, interactions, and tasks of a solution are proposed using industrial data collected on collaboration. These models should inform development of support to assist knowledge processing to improve work performance of stakeholders and consequent quality of outcomes.

1 Introduction

Quality innovation in reduced cost and time to market is critical for survival of organizations within the global, competitive product development (PD) context. Collaborative design has the potential in enabling organizations to achieve this. A design process is termed collaborative when a product is designed through the

---

Gokula Vijaykumar A.V. • A. Chakrabarti (✉)
Innovation, Design Study and Sustainability Laboratory (Ideas Lab), Centre for Product Design and Manufacturing, Indian Institute of Science, Bangalore 560012, India
e-mail: g.v.annamalaivasantha@cranfield.ac.uk; ac123@cpdm.iisc.ernet.in

S.P. MacGregor and T. Carleton (eds.), *Sustaining Innovation: Collaboration Models for a Complex World*, Innovation, Technology, and Knowledge Management, DOI 10.1007/978-1-4614-2077-4_5,

collective effort of many designers, including tools (Wang et al. 2002). Collaborative design is particularly relevant since designers currently expend a substantial proportion of their time to satisfy their knowledge needs. MacGregor et al. (2001) point out that engineers perceive 34% of their time is taken in sourcing and locating relevant information, and the mode of communication is 53% asynchronous with the rest synchronous, which demonstrates the importance of collaboration in knowledge processes.

Understanding collaboration among stakeholders (such as designers, manufacturers, managers, and consultants), the tools used, and the knowledge processes for designing should help organizations assess the quality of its collaboration and knowledge processes. This understanding, in turn, should help develop support to assist knowledge processes, so as to improve the work performance of stakeholders and consequent quality. Existing models for understanding knowledge processes through collaborative activities are inadequate in describing significant details of these processes; importantly, they do not stress the centrality of interactions in processing knowledge.

A collaborative model called Knowledge–Requirements–Interactions–Tasks, or "KRIT," which emphasizes interactions among designers, other stakeholders, and tools in processing knowledge during PD, has been developed. This model is used to understand how collaboration takes place among stakeholders in PD, from the point of view of knowledge processing. In the KRIT model, interaction plays the central role in identifying knowledge, requirements, and tasks with which requirements are addressed. The interactions cascade together to form a map of interactions, providing a view of all collaborations, and informing their quality from a knowledge-processing viewpoint.

In this chapter, the objective is to understand how knowledge, requirements, interactions, and tasks are satisfied during PD. A model of how satisfaction of requirements, tasks, interactions, and knowledge of solutions influence one another is also proposed, and discussed using data collected from two observational studies conducted in two organizations in India. The two organizations are chosen to highlight how a small, private enterprise and a medium, public enterprise contrast with each other in terms of their knowledge processing profiles. We detail the understanding obtained through these industrial case studies on the elements modeled in the two proposed models: the KRIT model and the Influence model. This understanding emphasizes the importance of collaboration and how the proposed models could be used to help sustain innovation through collaboration.

2 Innovation and Collaborative Design

The argument that collaboration and supporting knowledge processes enrich designers' creativity to produce innovative outcomes is highlighted widely in the literature. Larsson et al. (2003) observed that one-on-one conversations, held in parallel to a main discussion, were common in colocated teamwork and were a natural part of creative teamwork. Frankenberger and Badke-Schaub (1999) argued that availability of information is central to the success of design. They observed that designers spend more time individually than in teams, but critical situations occur during collaborations. Moritz and Schregenberger (1997) argued that prerequisites for producing creative solutions through cooperation are openness toward new ideas and viewpoints, application of efficient and effective state of the art methods, eventual objectives, and accumulation and distribution of information. Leonard-Barton and Sensiper (1998) posited that creative cooperation for merging knowledge from diverse disciplinary and personal skills-based perspectives is crucial to creating innovative, complex systems and products. Similarly, Sonnenwald (1996) found that design teams increasingly included participants from different domains to explore and integrate their specialized knowledge to create innovative and competitive artifacts and reduce development costs. Haymaker et al. (2000) considered approaches to collaborative design for new means of generating coherence and innovation by reformulating construction and flow of information. Lahti et al. (2004) pointed that computer supported collaborative environments for knowledge building provide a promising innovation to facilitate teamwork among designers, while Petre (2004) noted that innovative engineers are "hungry" for input, and work actively to maintain and update their knowledge base.

In the high-tech sector, knowledge is considered to be the only meaningful economic resource (Buchanan and Gibb 1998). MacMorrow (2001) argued that potential benefits of managing knowledge range from improving productivity, decision making, customer service, and innovation. Newell et al. (2002) argued that knowledge is used to support innovation within both teams and companies, and Cheung et al. (2008) demonstrated that knowledge reuse resulting from a repository type of knowledge management system actually inhibits creative performance of individuals, especially on the qualitative dimension. They argued that knowledge reuse for innovation focuses on knowledge integration through which others' knowledge is integrated into one's existing knowledge stock to accomplish an innovative task. These results emphasize the importance of collaboration and the associated knowledge processes.

3 Knowledge Processes Models

One common limitation in current literature in this area is that the concepts used are rarely defined in a systematic manner. The following definitions are used in this chapter to understand knowledge processes.

- *A knowledge element is defined as an entity (building block) processed in the PD process.* For example, function, behavior, and structure are some of the high-level knowledge elements for a product description.
- *A knowledge process is defined as the process through which knowledge elements evolve in their life cycle.* For example, search, retrieve, generate, capture, store, share, and (re) use are some of the commonly observed knowledge processes.
- *An agent is defined as a perceptible object through which designing occurs.* For example, designer, customer, computer, and documents are some of the agents.
- *An interaction is defined as a mutual or reciprocal action or influence of agents.* For example, "designer working with computer," "two designers working with a computer," and "many designers interacting with each other" are some of the interactions.

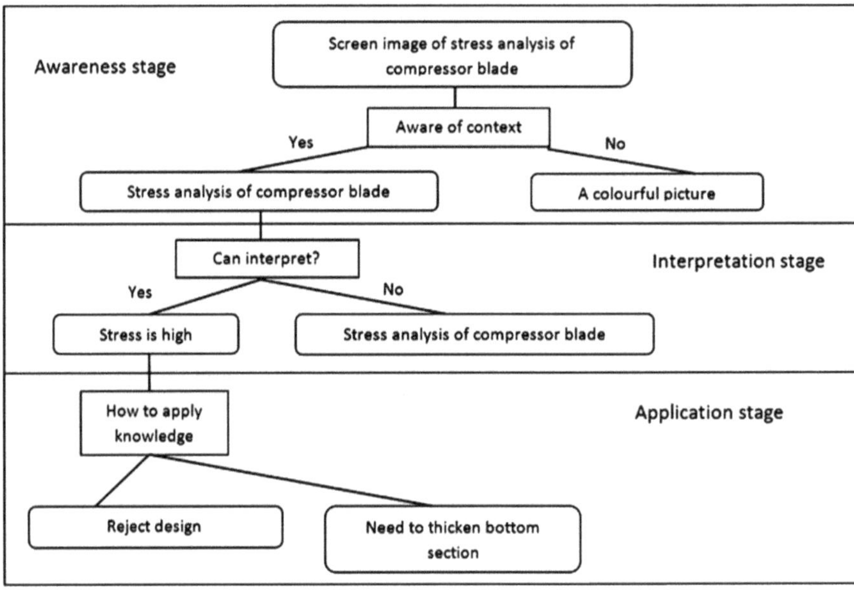

Fig. 5.1 Definition of data, information, and knowledge through stress analysis example (Ahmed et al. 1999)

The definitions of data, information, and knowledge are used from Ahmed et al. (1999). Data are taken as symbols or facts without context and are thus neither directly nor immediately meaningful. Information is data placed within some context. Knowledge is taken as a meaningful interpretation of information. We choose these definitions as they take into account scenarios in which a source and a user are involved in reciprocal actions. In our research, such scenarios are termed "interaction." Figure 5.1 explains these definitions through an example given by Ahmed et al. (1999).

To understand the knowledge processes involved in designing, many models have been proposed: these variously focus on the design process (French 1985; Pugh 1991), argumentation process (Kunz and Rittel 1970; MacLean et al. 1991), artifacts being designed (Chandrasekaran et al. 1993; Chakrabarti et al 2005), and activities of designers (Blessing 1994; Nidamarthi 1999). All these models provide rich descriptions in their own segments. However, for understanding the day-to-day knowledge processes of designers, we felt that the following points, missing in the current models, need to be incorporated.

- Interactions must be centered on the knowledge processes.
- Types of knowledge processes (e.g., knowledge capture, reuse) must be explicitly mentioned.
- Interlinks among knowledge elements (i.e., product and process aspects) must be highlighted and represented.

McDermott (1999) noted that approaches and tools developed to assist designers are inadequately adopted in industries due to an inadequate understanding of the knowledge processes of designers and industries. Analyses of relevant literature and observational data from industry indicate that the following points must be taken into account in any model to be used to understand the knowledge processes of designers:

- Each major knowledge element should be possible to be shown explicitly, with links to other knowledge elements.
- The interactions responsible in processing knowledge should be made explicit, and linked to associated knowledge elements.
- The model should provide a simple, easy-to-use, and meaningful representation for day-to-day knowledge processes.
- All major knowledge processes should be possible to be shown explicitly.

Given the points above, developing a new model is necessary. It should help organizations and designers to understand the dynamics involved in knowledge processing during PD. This, we argue, should help understand the associated knowledge processes, i.e., what knowledge is generated, captured and reused, and how (well) these are (currently) carried out. Besides being useful in aiding generation, capture and reuse of knowledge, this model should help provide insight to the process of knowledge transfer in an organization, which can be quite complex, requiring much time and effort to understand and assess. In this work, our aim is to support practice to better understand its collaborative knowledge processes.

4 Focus and Approach

The main foci of this chapter are to understand how knowledge is processed during collaborative PD in industry, and how efficient these processes are. We developed two models: the KRIT model, and based on this, an Influence model, to address these.

The KRIT model helps understand the knowledge processes during collaborative PD. Its distinguishing feature is the central role played by interactions in knowledge processing, something not explicitly taken into account in earlier models. Interaction of designers with other people and tools, we argue, is the vehicle through which knowledge processing occurs during PD. Our primary hypothesis in this model is the mutual influence of interactions and knowledge processes on one another. To understand how efficient these knowledge processes are, an Influence model has been developed with the KRIT model as the basis. These two models are detailed in Sects. 5 and 6.

To realize the benefits of these proposed models, two ethnographic observational industrial studies were undertaken: one in a small, private enterprise (providing innovative solutions and services in consumer products) termed SmallCADCo; and the other in a medium R&D organization (developing special purpose aircraft) termed MediumAeroCo. SmallCADCo is a joint (50:50) venture between a reputed academic institution and an IT company. The organization is 12 years old and consists of 15 employees. A substantial number of interactions for the personnel in this organization occur in consulting domain experts from the academic institution. MediumAeroCo collaborates with various public and private sector companies and

Table 5.1 Characteristics of companies involved

Characteristics	SmallCADCo	MediumAeroCo
Joint ventures	Partnership between private and academic institution	Public organization (funded by Government of India)
Number of employees and groups	Less and no specific groups	Medium and more groups
Complexity of products	Less and Medium	Highly complex and integrated
Place of work	Colocated	Distributed around India

academic institutions to design and develop special purpose aircraft. The groups in the organization are structured as: Systems Directorate, Propulsion Systems, General Systems, Air Frame, Flight Test, Integrated Flight Control System, Quality Assurance and System Effectiveness Group, Independent Validation and Verification, Project Management, Aerodynamics Research and Development, Protovehicle and Productionisation, Advanced Projects and Technologies, Information Systems, and other administrative departments. A total of around 240 employees are distributed across these groups. Both the organizations observed primarily serve Indian markets. The specific characteristics of these organizations that might influence collaboration and networking are summarized in Table 5.1.

In SmallCADCo, three designers involved in different projects were observed serially for 3–7 days each. The designers observed were novices with 1–3 years of work experience. All projects observed were carried out for the first time by these designers. The average duration observed per day for the subjects were 5.4, 3.0, and 2.8 h. In MediumAeroCo, seven designers were observed with 1–40 years of work experience. All except one designer were at senior levels in the organization. The observed number of days varied from 9 to 27. The average duration observed per day for each designer was 4.6, 2.7, 3.5, 1.8, 1.3, 2.3, and 3.3 h. Different projects involved in different stages of PD were chosen to evaluate the general applicability of the models proposed.

Data was collected using questionnaires, data sheets, voice recordings, and unstructured interviews, on the following topics: purpose of the tasks, interactions, place and duration of interactions, whether interactions were satisfying or not, project details, and subjects involved in the observations. All subjects observed informed that the observations had not disturbed or influenced their activities. Though we focused only on a total of ten designers in the two organizations, the data collected also included all other designers who interacted with these core ten designers during the observational period.

5 The KRIT Model

We propose the Knowledge–Requirements–Interactions–Tasks (KRIT) model, in which interactions of designers with people and tools are central to processing knowledge during PD. We argue that interactions lead to various knowledge elements, and

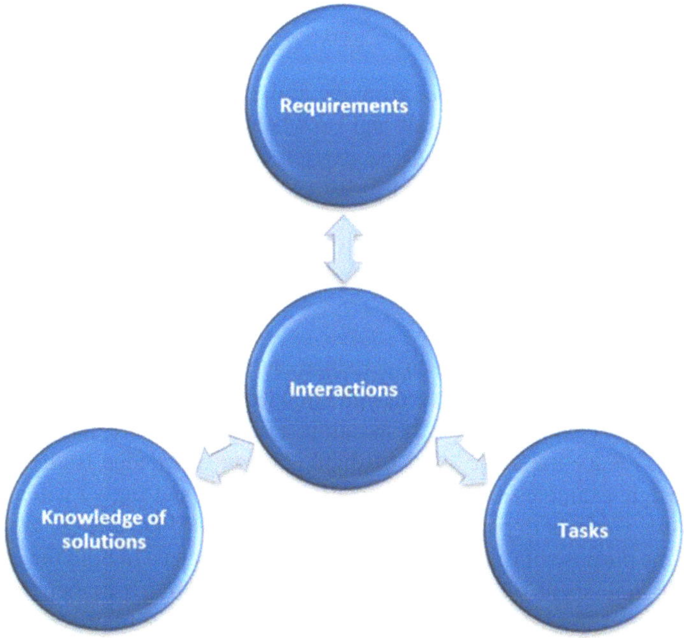

Fig. 5.2 Links between knowledge elements and interactions (the KRIT model)

these knowledge elements lead to various, new interactions. Nonaka et al. (2000) have a similar hypothesis for knowledge creation. They state that organizations create knowledge through interactions between explicit and tacit knowledge.

In order to encompass knowledge elements from both product and process points of view, "requirements" (representing product knowledge) and "knowledge of solutions" (representing product and process knowledge) and "tasks" (representing process knowledge) are included in the model. Their definitions are as follows:

- *Requirements*: Intended aspects of the product considered by designers during PD. For instance: "What is the working hours mentioned for filter head?"
- *Knowledge of Solutions*: The outcomes, i.e., artifact being designed, produced by designers to satisfy requirements. For example: "Extra steel plate should be added here because there is a gap of 1 cm."
- *Tasks*: A piece of work to be done to satisfy requirements. Two examples are: "to modify existing mold design," and "to measure dimensions from physical model."

In order to provide insight into the knowledge processes, links among requirements, tasks, interactions, and knowledge of solutions are explicitly represented. Using interactions as the core enabler, links are established among the knowledge elements, see Fig. 5.2.

6 Observations in Industry

Analyses of information collected from the ten subjects show that all three knowledge elements (requirements, tasks, and knowledge of solutions) are present in the collaborations and are connected via explicit interactions. We now discuss the observations related to each element.

6.1 Knowledge of Solutions

Knowledge of solutions has been classified into "product-based" and "process-based" knowledge. Product-based design knowledge is concerned with the objects being designed; examples are "it blows air at a certain pressure" and "the function might be to reduce the noise." Process-based design knowledge is concerned with how to design; examples are "cut till the inside surface" and "now I will make it this way." In this work, both product-based and process-based knowledge are classified based on the purpose of the tasks carried out by the subjects. In both the studies, the amount of time spent on product-based knowledge is much higher than that on process-based knowledge. This indicates that irrespective of the complexity of products being designed, and size and number of groups within the organizations, the focus has primarily been on knowledge about the product.

6.2 Requirements

As classified by Nidamarthi (1999), two types of requirements: Solution-Neutral Requirements (SNRs) and Solution-Specific Requirements (SSRs) are observed in both the studies. SNR describe the generic requirements which designers address with their designs. Solution-Specific requirements are specific to certain solutions only. For example, in a project designing an injection mold for a given component, its manufacturability and strength are SNRs, whereas questions of "how to avoid liquid leakage due to this dwell?" and "how much length should be given such that the engravings should not be affected?" are SSRs. The amount of discussions around SNR and SSR is much higher in SmallCADCo than in MediumAeroCo, due to the longer design time involved in MediumAeroCo.

6.3 Interactions

Interactions are a primary constituent for knowledge processing in an organization. In SmallCADCo and MediumAeroCo, respectively, 19 and 17 different types of interaction were found to be present. These types are classified based on the variety and the number of agents participating in a single interaction. The variety and number of agents involved in these interactions demonstrate the complexity of collaboration within these organizations. In both studies, the agents involved in these interactions, apart from humans, were: computer, measuring device, prototype model, document, notebook, paper, calculator, and whiteboard. The designations of the humans involved were: designer, engineer, design student, academic, external consultant, manager, supplier, customer, manufacturer, and scientist. In both studies, the interactions that occurred most frequently are: "one designer working with a computer," "two designers working with a computer," and "two designers interacting with each other." In MediumAeroCo, two designers spent almost all their time individually interacting with a computer only. Tools for supporting knowledge capture and reuse must support these interactions, to ensure that capture and reuse can be built-in in a natural way into a designer's work patterns.

6.4 Tasks

We classified tasks into six categories, based on knowledge exchanges performed by the subjects:

1. Generating knowledge alone (by the subject)
2. Generating knowledge (by the subject) with others
3. Giving knowledge (by the subject) to others
4. Taking knowledge (by the subject) from others
5. Searching for knowledge (by the subject)in documents or computer
6. Capturing knowledge (by the subject) in documents or computer

The first two categories represent knowledge generation; the next three represent knowledge reuse, while the last one represents explicit knowledge capture; note that all six categories of tasks might involve implicit knowledge capture. The amount of time spent in each task varied substantially between novice and expert designers. The variations were greatest in *generating knowledge with others* and *giving knowledge to others*. Novice designers, understandably, spent more time in *generating knowledge with others* and less in *giving knowledge to others,* whereas expert designers spent most of their time in *generating knowledge alone*. The amount of time spent on tasks to capture knowledge was very low, in both the studies. Knowledge capture happened only as part of the other five knowledge exchanges. The reasons for this could be due to the time pressure of the projects, low awareness of the importance of knowledge reuse, and since most of the projects are perceived to be unique in nature, low incentives to capture knowledge due to the perception that chances of reuse is low. We argue that increasing awareness and possibility of knowledge transfer from one project to another would substantially improve the proportion of knowledge captured, and reduce the amount of time spent on giving and taking knowledge, both impacting on the amount of time of designers involved in these tasks. The average time spent on these knowledge exchanges in both the organizations are summarized in Table 5.2.

Table 5.2 Average time spent on knowledge exchanges in SmallCADCo and MediumAeroCo

Types of knowledge exchanges	SmallCADCo (%)	MediumAeroCo (%)
Generating knowledge alone	29.8	42.9
Generating knowledge with others	44.2	22.1
Taking knowledge from others	13.4	3.3
Giving knowledge to others	5	4.5
Searching for knowledge in documents or computer	5.9	7.8
Capturing knowledge	1.7	19.4

In MediumAeroCo, the only novice designer involved found it difficult to search the documents available. He spent less time in searching and taking knowledge from others. This designer had to repeat some tasks several times. More attention needs to be provided to understand resources, including experts, from which knowledge can be gained.

Informal capture in private notebooks was predominant in MediumAeroCo, with the drawback that no one else could access this knowledge. To overcome this, a method to share informal knowledge capture might be necessary. Expert designers were largely preoccupied with their own tasks, and rarely interacted with others to share knowledge. This isolation must be avoided in a meaningful manner to facilitate efficient transfer of expertise. Experts spent more time in searching for knowledge in documents, indicating that they found the documents more useful, and that experience played a vital role in identifying appropriate knowledge resources. It would be interesting to investigate what knowledge was used to identify the documents and search them.

By studying the variations in time spent across tasks by each designer, we found that designers stick to their preferred modes of working. Capturing knowledge in MediumAeroCo was forced through adoption of standards, only as required for standards accreditation before the inspection period. However, this was not part of the normal routine of the designers involved. This scenario needs to be changed to incorporate a practice of capture as part of the daily routine of designers. Relevance of the captured documents is assessed by comparing their content with the questions asked by designers during PD. This revealed that only 18.7% of the answers to the questions asked were captured in, and therefore possible to be answered using, the documents; this leaves substantial scope for improving knowledge capture.

In SmallCADCo and MediumAeroCo, respectively, 18.4% and 7.8% of the time were spent in taking and giving knowledge to/from others, which is less than the 20–30% reported in literature (Court and Culley 1995; Marsh 1997; MacGregor et al. 2001). This decrease could be due to the post-social web revolution. The percentage is less in MediumAeroCo, possibly due to the greater experience of its designers observed. The variations between SmallCADCo and MediumAeroCo, and the fact that each designer stuck to his preferred working pattern, illustrate that knowledge processes should be supported in a more personalized manner, while utilizing organizational resources effectively. These observations emphasize the importance of collaboration among agents in the various knowledge exchanges and also highlight the importance of solo work within any framework of collaboration: collaboration always includes both individual work and team interactions put together to create a harmonious whole.

7 Influence Model and Assessment

Using the KRIT model, the PD process is explained as follows. Requirements are taken as the primary objectives to be fulfilled in PD. Each requirement requires some tasks with purposes and outcomes (knowledge of solutions) to be carried out, which are generated through various interactions. As a result of these interactions, knowledge is generated to satisfy the tasks and fulfill the requirement. Each task has a set of knowledge to be processed, which are processed through a complex variety of interactions. The knowledge generated to satisfy the tasks will be input for other tasks and requirements and will be carried out further down the PD process.

Figure 5.3 illustrates the Influence model developed from the KRIT model. Requirements satisfaction is the primary objective to be fulfilled. To satisfy each requirement, a set of tasks with purposes and outcomes should be satisfied. A set of knowledge of solutions have to be processed and satisfied to complete a task. Knowledge of solutions could be satisfied only if a set of interactions among designers (and tools) gets satisfied. The following subsections define the four satisfaction levels, and observations from SmallCADCo. We restricted the analyses to SmallCADCo due to the more detailed information collected in this study.

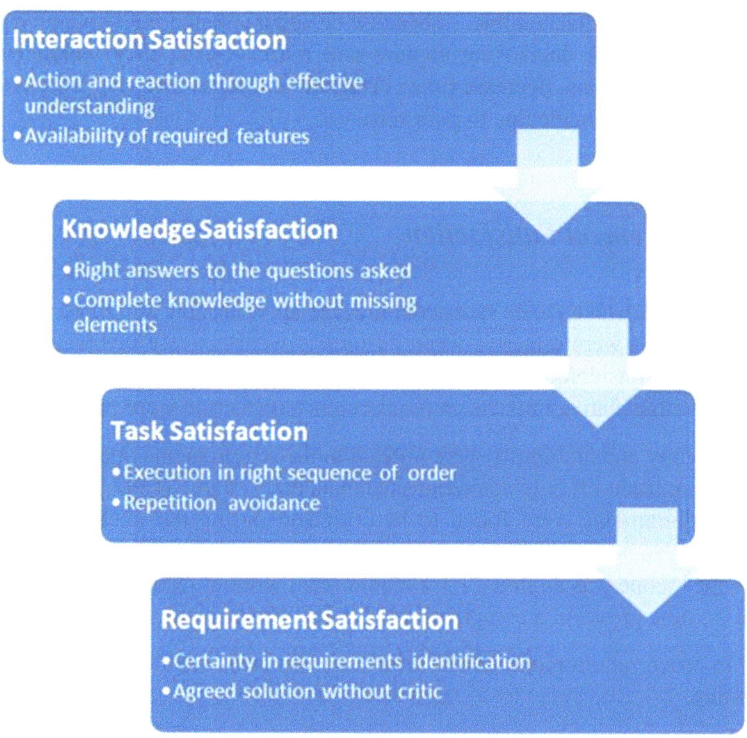

Fig. 5.3 Influence model by hierarchy tree of element satisfaction

7.1 Knowledge Satisfaction

We take knowledge to be satisfied, if right answers are obtained for the questions asked. In addition, knowledge satisfaction is high, if the process follows a Generate–Evaluate–Select cycle. This is based on the observation that designers were highly satisfied, when a proposed solution was accessed, justified and agreed commonly among the stakeholders involved. Knowledge satisfaction was the least when:

- The questions asked were not answered.
- The answers were refuted or solutions criticized.
- Differing points of view existed across stakeholders.
- Complexity of solutions was high.
- Answers were incomplete, missing or were unknown.
- Assumptions were made without proper verification.

Such instances should be prevented from happening to increase knowledge satisfaction in PD. A detailed study on the questions asked by the designers in SmallCADCo revealed that nearly 50% of the old queries were answered by colleagues (Vijaykumar and Chakrabarti 2008). This would significantly impact design time, as each such interaction consumes time of both the designers and the colleagues with whom they interact. This was either because the captured documents were inadequate or inaccessible, or because designers trusted their colleagues more than the documents. Increasing capture and retrieval efficiency would enhance knowledge satisfaction, decrease time consumed in discussions, and prevent unnecessary assumptions made due to poor retrieval.

7.2 Requirement Satisfaction

Requirements are taken to be satisfied if they are appropriately identified and solved. These processes were effective when customers were actively involved, life cycle phases were considered, and needs behind requirements were recognized. The instances with negative impacts on requirement satisfaction were the following:

- Uncertainty and ambiguity were noticed in the requirements chosen.
- Immature trade-off between requirements at early stages of PD was found.
- The requirements were found to be criticized during the later, more detailed stages.
- A wider scope was assumed for a requirement without justification.
- Due to time pressure, a compromised decision was taken to satisfy a requirement.

To improve requirement satisfaction, such instances should be prevented from happening.

7.3 Interaction Satisfaction

As defined before, interaction is the mutual or reciprocal action or influence of agents. Before defining interaction satisfaction, we clarify what is a single interaction Is. A period of observation is considered a single interaction if during that period the goal has not changed, the outcome is not achieved and the agents are not changed. By analyzing the interactions observed in SmallCADCo, the instances that negatively impacted the interactions are noted. Minneman (1991) enlists the various ways by which design outcomes emerge from interactions among individuals and groups as they establish, develop, and maintain a shared understanding. Negotiating understandings, conserving ambiguity, tailoring engineering communication for recipients, and manipulating mundane representations are identified as some of the crucial group activities. The following are identified as negative instances:

- Difficulty in visualizing articulations and features were noted.
- Misidentification of features was identified.
- Substantial time was spent in establishing common understanding. Misinterpretations were noted, time was spent in clarifying and in creating awareness to maintain a common understanding among subjects.
- Avoidance of communication was noted with some stakeholders due to fear of time consumption.
- No common software was used among all the stakeholders. Interoperability between software was an issue. It was difficult to use files across different software of the same type, e.g., CAD software.
- Identification of documents was time consuming. Tracing their locations was difficult.
- The size of the computer files made document sharing via e-mail difficult.
- Some of the required features were unavailable in some of the software used.
- Some of the required software was unavailable.
- Some of the interactions with stakeholders had been delayed or postponed due to unavailability of the stakeholders.
- The place of work was not tidy; documents were placed awkwardly and made discussion difficult.

From these observed instances, we argue that interaction will be satisfied if intended actions and reactions take place through required composition and capabilities of people, tools, process, and information (Fig. 5.4). The current situations in the observed organizations should be improved substantially by considering these factors, to satisfy interactions.

7.4 Task Satisfaction

A task is taken to be satisfied if another task, dependent on this task, is found to be carried out subsequently in the PD process. Tasks chosen based on customer preferences led to positive structuring of tasks. Tasks were found to be executed in an ad

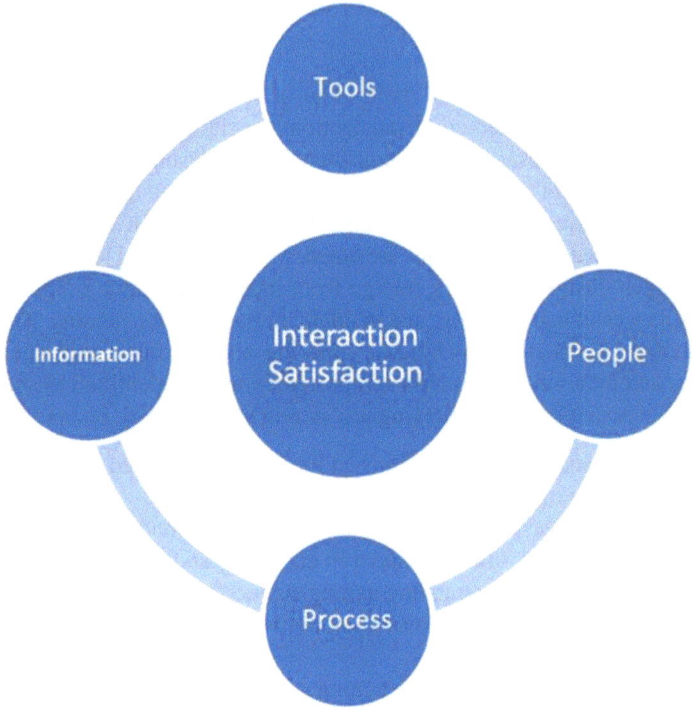

Fig. 5.4 Common merge of factors influencing Interaction Satisfaction

hoc manner, without following any formal structure. Tasks were carried out in an opportunistic, subjective fashion. Often, designers carried out new tasks without completing current tasks. This behavior led to task failure, task repetition, and iterations. The scope of tasks was sometimes reduced, and some tasks were removed due to the perceived effort and time involved in executing them. Difficulty to plan tasks and schedule timings was also observed. Assessment criteria should be modified to stress quality of the tasks' outcomes. Tasks should be executed in the right sequence to enable more effective task satisfaction, as repetitions could be avoided.

8 Discussion and Conclusions

The overall aim of this work is to support industry develop high-quality novel designs in reduced time through effective knowledge processes. Using the proposed KRIT model and Influence model, the understanding obtained on the knowledge processes involved in collaborative PD from two industrial studies is reported. This understanding should help improve collaborative capabilities of organizations, which is important for improved innovation in challenging business markets.

The KRIT model highlights the centrality of interactions in knowledge processing, something not adequately highlighted in earlier models. A major, potential benefit of this model is in representing knowledge processing in terms of interactions to develop requirements, tasks, and knowledge of solutions. These results should inform development of support to assist knowledge processes to improve work performance of stakeholders, and consequent quality. The model makes explicit and highlights the various stakeholders involved in these processes such as designers, engineers, and external consultants. The KRIT model is primarily developed to understand and support designers and industries. However, this model could also be used to study in detail the effectiveness of collaborations among quadruple helix actors (i.e., collaborations among firms, users, public organizations, and universities).

The Influence model proposed is intended to help assess the quality of collaborations and knowledge processes in PD. We argue that understanding the degree of satisfaction of interactions, knowledge, task, and requirement, collaborations should be possible to be assessed. A list of potential issues for each of these has been identified. Issues involved in data, information, and knowledge transformation are highlighted in interactions satisfaction through difficulties incurred due to lack of awareness and misinterpretation. The results indicate that substantial enhancement in knowledge processing is possible if the interactions carried out by designers during PD could be improved. Improvement in interactions and its impact on knowledge processing, however, need to be studied in detail, to provide a theoretical basis on which strategies for effective knowledge processing could be developed.

References

Ahmed S, Blessing L, and Wallace K (1999) The relationships between data, information and knowledge based on a preliminary study of engineering designers. ASME Design Theory and Methodology Conference, Las Vegas, Nevada.

Blessing L (1994) A process-based approach to computer-supported engineering design. Ph.D. Thesis, University of Twente, Netherlands.

Buchanan S, and Gibb F (1998) The Information audit: An integrated strategic approach. Int J Inf Manage 18: I, 29–47.

Chakrabarti A, Sarkar P, Leelavathamma B, and Nataraju B.S (2005) A Functional Representation for Aiding Biomimetic and Artificial Inspiration of New Ideas. Artif Intell Eng Des Anal Manuf 19:2 113–132.

Chandrasekaran B, Goel A, and Iwasaki Y (1993) Functional Representation as Design Rationale. IEEE Computer 26 48-56.

Cheung P. Chau P.Y.K. and Au A.K.K. (2008) Does knowledge reuse make a creative person more creative? Decision Support Systems 45: 219–227.

Court A.W. and Culley S.J (1995) A methodology for analyzing the information accessing methods of engineering designers. Proc. of Int. Conf. on Engineering Design 2: 523–528.

Frankenberger E, and Badke-Schaub P (1999) Information management in engineering design – empirical results from investigations in industry. Proc of Int Conf on Engineering Design 911–916.

French M. (1985) Conceptual Design for Engineers. Design Council – Springer Verlag, London, UK.

Gokula Vijaykumar A.V, and Chakrabarti A (2008), Understanding the Knowledge Needs of Designers During Design Process in Industry. J Comput Inf Sci Eng 8 011004-1–011004-9.

Haymaker J, Keel P, Ackermann E, and Porter W (2000) Filter mediated design: generating coherence in collaborative design. Design Studies 21: 205–220.
Kunz W, and Rittel W (1970) Issues as Elements of Information Systems. University of California Working Paper No. 131.
Lahti H, Seitamaa-Hakkarainen P, and Hakkarainen K (2004) Collaboration patterns in computer supported collaborative designing. Design Studies 25: 351–371.
Larsson A, Torlind P, Karlsson L, Mabogunje A, Leifer L, Larsson T, and Elfstrom B-O (2003) Distributed Team Innovation- A framework for distributed product development. Proc of Int Conf on Engineering Design.
Leonard-Barton D. and Sensiper S (1998) The role of tacit knowledge in group innovation. California Management Review 40:3.
MacGregor SP, Thomson AI, Juster NP, (2001) Information sharing within a distributed collaborative design process: A case study, Proceedings of DETC'01.
MacLean A, Young R, Belloti V, and Moran T (1991) Questions, Options, and Criteria: Elements of Design Space Analysis. Human-Computer Interaction 6:3–4 201–250.
MacMorrow N (2001) Knowledge management: an introduction. Annual Review of Information Science and Technology, Williams ME, Ed. 381-422. Medford, NJ: Information Today.
Marsh J R, (1997) The capture and utilisation of experience in engineering design. Ph.D. Thesis, St. John's College, Department of Engineering, University of Cambridge.
McDermott R, (1999) Why information technology inspired but cannot deliver knowledge management. California Management Review 41:4 103–117.
Minneman S L (1991) The social construction of a technical reality: Empirical studies of group engineering design practice. Ph.D. Thesis, Department of Mechanical Engineering, Stanford University.
Moritz E F, and Schregenberger J W (1997). How to achieve creative synergy – Contemplations, cases, and concepts in cooperative design, Proc of Int Conf on Engineering Design 2: 149–156.
Newell S, Robertson M, and Scarborough H (2002) Managing Knowledge Work. Palgrave, London.
Nidamarthi S (1999) Understanding and Supporting requirement satisfaction in the design process. Ph.D. Thesis, Gonville and Caius College, University of Cambridge.
Nonaka I, Toyama R, and Nagata A (2000) A Firm as a Knowledge-creating Entity: A New Perspective on the Theory of the Firm. Industrial and Corporate Change, 9:1.
Petre M (2004) How expert engineering teams use disciplines of innovation, Design Studies 25 477–493.
Pugh S (1991) Total Design : Integrated Methods for Successful Product Engineering, Addison Wesley, UK.
Sonnenwald D H (1996) Communication roles that support collaboration during the design process. Design Studies 17: 277–301.
Wang L, Shen W, Xie H, Neelamkavil J, and Pardasani A (2002) Collaborative conceptual design – state of the art and future trends, Computer-Aided Design 34: 981–996.

Chapter 6
Public Policy Support to Triple Helix R&D Collaborations: A European Model for Fourth Pillar Organizations

Alexandra Simon and Pilar Marquès

Abstract Fourth pillar organization is the name used for independent, not-for-profit, member-based organizations that combine funding from the government and the private sector and are conceived to facilitate the complex collaboration among triple helix participants. This chapter sets out to explore how these organizations are effective tools for governments to boost collaborative innovation. It analyzes four cases of successful fourth pillar organizations in three different European countries—namely, Holland, Spain, and Sweden—and uncovers a different model of such organizations to the one found in previous research for Canada and the USA. Particularly, the government has a more proactive and preeminent role, as well as notable participation of the private financial sector. We also found that fourth pillar organizations tend to complement existing industry and R&D capabilities so that they have a more limited or extended role depending on the strengths and weaknesses of the existing triple helix system. This chapter further contributes to understanding better why fourth pillar organizations have been created and how they can contribute to facilitate triple helix collaboration. It, therefore, provides ideas for reflection so that government and industry can better guide their future action and commitments.

1 Introduction

In a remarkably short time, economic globalization has changed the world's economic order, bringing new challenges and opportunities. Europe cannot compete in this new environment unless it becomes more innovative and responds more

A. Simon • P. Marquès (✉)
Department of Business Organization, Universitat de Girona,
Campus Montilivi, Girona 17071, Spain
e-mail: alexandra.simon@udg.edu; pilar.marques@udg.edu

S.P. MacGregor and T. Carleton (eds.), *Sustaining Innovation: Collaboration Models for a Complex World*, Innovation, Technology, and Knowledge Management,
DOI 10.1007/978-1-4614-2077-4_6, © Springer Science+Business Media, LLC 2012

effectively to consumers' needs and preferences (European Commission 2009). Innovation in products, services, processes, and the formation of new business enterprises is crucially important to every economy, and has long been a main concern of the European Union (EU) policymakers (Audretsch et al. 2009). It is essential that authorities at all levels—local, regional, national, and supranational—contribute to creating innovation-friendly environments for their industry, despite the current constraints on public budgets. It is, therefore, relevant to optimize the efforts of governments to improve innovation policies, as well to provide new mechanisms of effectively managing knowledge transfer and innovation implementation among the different partners involved in the innovation process. Fourth pillar organizations, which constitute the topic of this paper, have become an important tool in the successful realization of these goals (Dalziel 2005; Johnson 2008).

The purpose of innovation policy should be to create a favorable environment and framework within which individuals and firms are encouraged to steadily improve technological products, processes, and practices. According to Garofoli and Musyck (2001, 2003), the key precondition for highly innovative processes and outcomes within a regional context is not necessarily investment in the knowledge-producing sector. Instead, a major requirement is the strengthening of networks as well as of agents of change, but also of organizational and institutional patterns, with the goal of improving the environmental conditions for innovators and entrepreneurs at the local level. Within a regional innovation system, the performance of individuals and firms is a function of the regional conditions (Cooke et al. 2004; Musyck and Reid 2007).

Fourth pillar organizations are defined as independent, not-for-profit, member-based organizations that provide a facilitating role among the three traditional pillars in our economy: industry, universities and other higher education institutions, and government. Fourth pillar organizations leverage private and public investment to implement activities, such as shared-cost R&D programs, build shared R&D infrastructure, and supply technical products and services (Liljemark 2004). Since they are promoted and funded—at least partially—by the public sector, fourth pillar organizations are not nongovernmental organizations (NGOs).

An example of a fourth pillar organization is Precarn in Canada, originally conceived as an industry-led consortium with the primary mission to support industry-relevant, market-oriented collaborative R&D in the sector of intelligent systems (Johnson 2009). Intelligent systems consist of technologies based on artificial intelligence and computer simulation systems. The Precarn team manages a program that provides resources, such as financial support and managerial oversight on sponsored triple helix projects. The actual source of funding comes both from private and government sources, with the aim of leveraging public money such that both public and private sectors share the costs of innovation.

This paper is framed within a European project called "Creating Local Innovations for SMEs through a Quadruple Helix" (CLIQ). The quadruple helix model argues that, in addition to the three pillars of the triple helix model, civil society needs to be incorporated into the process of knowledge creation (Carayannis and Campbell 2009; MacGregor et al. 2010). The long-term aim of CLIQ is to optimize the benefits of globalization and innovation for SMEs and entrepreneurs in medium-sized towns, with the main objective to strengthen local authority policy and their capacity to support innovation more effectively. Despite the project's focus on the quadruple helix, participants still highlighted the need for better coordination within the triple helix collaboration that partners had already put in place. They acknowledged the potential of fourth pillar organizations as a good way to coordinate triple helix systems, and we found some successful fourth pillar organizations within the scope of the CLIQ project.

Given this context, the main objective of this paper is to analyze how fourth pillar organizations are created, identify the role of public authorities, and describe the role of the fourth pillar organizations within the triple helix system. The focus is given to the study of the government's authority policy and its capacity to support innovation more effectively via this type of organization.

There are already some analyses about fourth pillar organizations, but they are geographically limited to Canada (Liljemark 2004; Dalziel 2005; Johnson 2008, 2009). It is, thus, interesting to continue the analysis of fourth pillar organizations and compare whether the findings for Canada are comparable to the existing fourth pillars in Europe.

We begin by discussing the overall framework of European innovation policy. The next section discusses the support that fourth pillar organizations can receive from governments in the efforts to facilitate the process of innovation and technology commercialization in triple helix environments. The chapter then details the methodology used and results obtained, presented as case-level descriptions and analysis, with particular reference to the different structural models found in these European cases. A final conclusion section assesses the main findings and contribution of our research.

2 Innovation Policy in the European Union

The relevant framework for European innovation policy is the vision of *Europe 2020*, launched in 2010, to replace the mostly failed Lisbon agenda. This strategic plan is meant to help Europe overcome the global economic crisis and recover lost ground from the previous strategic vision. The three priorities of the Europe 2020 strategy are:

- Smart growth, developing an economy based on knowledge and innovation
- Sustainable growth, promoting a more resource efficient, greener and more competitive economy
- Inclusive growth, fostering a high-employment economy delivering social and territorial cohesion

Within this vision are a series of top-level targets, such as the Lisbon target of spending 3% of the EU GDP on R&D, this time aimed for by 2020, which could create 3.7 million jobs and increase annual GDP by close to €800 billion by 2025.

The core of the Europe 2020 action to achieve such top-level targets is represented in a total of seven flagship initiatives across the three priorities.

One of these flagship initiatives is the *Innovation Union*, launched also in 2010, designed to contribute to smart growth. By raising arguments for a more strategic approach to innovation, the aims of this initiative are to boost Europe's research and innovation performance by speeding up the process from ideas to markets. For this endeavor, the Innovation Union presents a set of requirements, including:

- The need to continue to invest in education, R&D, innovation, and information and communication technologies (ICTs)
- To be carefully protected from budget cuts
- Increased integration and performance of the EU and national research and innovation systems
- Mobilizing knowledge across Europe by means of the completion of the European Research Area
- Better SME access to EU programs, promoting smart regional specialization strategies
- The need to get more innovation out of research by means of enhanced cooperation between the world of science and business
- Removing barriers for entrepreneurs to bring ideas to the market
- Launching European innovation partnerships to accelerate research, development, and market deployment of innovation, especially in areas of concern for citizens, such as climate change, energy efficiency, and healthy living

Another flagship initiative is focused on competitiveness for sustainable growth: an industrial policy for the globalization era, launched in October 2010, which details the measures necessary to fully exploit the European market of 500 million consumers and 20 million entrepreneurs. Within the measures proposed, innovation performance is addressed through actions in sectors, such as advanced manufacturing technologies, construction, biofuels, and road and rail transport, particularly in view of improving resource efficiency.

Innovation policies are currently being implemented or considered in many EU countries, introducing stimulus packages, involving actions to foster research and entrepreneurship, and giving support to intermediate organizations that help in the innovation process, as well as investment in infrastructure, including ICT networks, human capital, and green technologies (Stark and Wolf 2007).

3 Public Policy Support to Fourth Pillar Organizations Managing Triple Helix R&D Collaborations

Governments have the mandate to increase economic and social well-being, national security, and administrative efficiency. Knowledge is an input to economic growth and social development, and governments seek to promote the generation of knowledge and its application to the economy, that is, innovation. In order to implement these

Table 6.1 Government innovation policy options

	Direct interventions	Indirect interventions
Directed R&D	Government laboratories, intermediate organizations (fourth pillar organizations)	Research grants to universities and firms
General R&D support	Technology-based projects	R&D tax credits
Directed science and technology (S&T) activities	Testing, standards, data collection	Regulatory activities
General S&T support	Technology outreach	Science and technology education

Source: Holbrook (2002)

goals, governments have a variety of innovation policy options, including both direct and indirect actions, as presented in Table 6.1. According to this classification, fourth pillar organizations are a direct intervention that governments can use to direct innovation to the desired sectors or typologies.

The incentives to innovate are different in the private than in the public sector. Innovation in the private sector can result in large financial rewards and greater market share, and thus be attractive to managers and other employees with innovation-oriented rewards systems. However, in the public sector, there is no such context. It is very likely that the possible financial rewards of innovation do not transcend to the individuals and teams involved in the innovation but go instead to the state. And since the public sector has traditionally been a monopoly provider of some goods and services, people in the public sector have had little incentive to engage in innovation. Therefore, it is interesting to consider how innovation can be leveraged by means of engaging the private and public sectors together. Fourth pillar organizations can solve this need as they involve both public (government, academia) and private (firms) actors in the realization of R&D projects.

Fourth pillar organizations are considered a vital tool for governments wishing to strategically invest in the development of new technologies, and their contribution to the development of innovation and wealth across all industrial sectors is a basic requirement for countries. They accelerate product development and ensure a faster time to market for participating companies. They can help produce higher quality products and services and increase production of high-value, high-knowledge components of many export commodities. They create jobs, develop new expertise, and build multidisciplinary teams to drive breakthrough research and discovery (Canarie et al. 2003).

Fourth pillar organizations constitute the ideal governance structure for the management of collaborative R&D projects directed to the technology transfer efforts and innovation strategy of a government. Figure 6.1 depicts the typical structure of a fourth pillar organization project, adopted from Johnson (2008), which always involves partners of at least three types of organizations: technology developers, technology users, and academic partners.

Fourth pillar organizations that create, manage, and regulate innovation are important strategic mechanisms that can be used to build the technological infrastructure of a country. They need to be closely linked with their government, one of their main supporters. This is important because, without government support, the

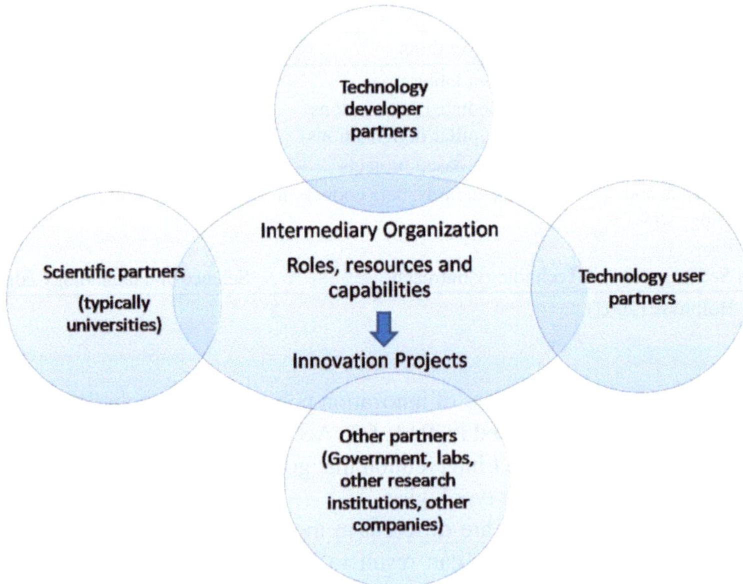

Fig. 6.1 Fourth pillar organizational structure and project model for supporting triple helix collaborative R&D projects (Source: Johnson 2008)

science and technology (S&T) benefits for the region or country may not materialize (Johnson 2009). Therefore, fourth pillars constitute an efficient tool for governments to expand their innovation policies without becoming directly involved in the processes of legitimating the technical merit of R&D projects or allocating funds for a triple helix partnership. Instead, this can be dealt directly by the fourth pillar organization (Johnson 2008).

4 Empirical Method and Case Selection

The research on which this chapter draws involves four case studies of fourth pillar organizations from different European countries: one from Spain, one from the Netherlands, and two from Sweden. The case studies are an illustration of the organizations' approach to managing multi-actor R&D projects effectively.

In this research, a case study is defined as "an empirical inquiry that investigates a contemporary phenomenon within its real-life context, when the boundaries between phenomenon and context are not clearly evident, and in which multiple sources of evidence are used" (Yin 1989, p. 23). For example, the research described here examines how fourth pillar organizations can help governments to successfully transfer technology among the different actors in triple helix partnerships. Such an approach is useful in exploratory modes of research and can provide detailed understanding of particular situations which may then be utilized inductively to create better theory, in this case how to manage European triple helix organizational collaboration.

Table 6.2 Characteristics of fourth pillar organizations

Fourth pillar organizations	
Type:	Funding:
Independent	Private funding
Member based	Public funding
Non for profit	
Partners:	Main purpose:
Industry partners	Implement shared-cost R&D programs
Academic partners	Build shared R&D infrastructure, supply technical products and services
Government partners	

The case studies are based on two main sources of information. First, secondary information was provided by the organizations, such as annual reports and other publicly available documents, including Web pages and academic articles. Second, primary information was gathered using a semistructured questionnaire targeting the directors of the organizations with the following six questions:

- How does your organization work internally?
- How does your organization work externally? How does the network of partners work?
- How do you understand success within your organization?
- What are the critical success factors for your organization?
- How does your organization facilitate triple helix collaboration?
- What are the main roles of your organization?

Fieldwork was a fundamental part of this investigation. The initial contact was established by means of personal visits during the CLIQ project, followed up by telephone and e-mail, starting in March 2009 and finalizing in August 2009. The organizations were asked to read the case study, which had previously been written using secondary data, to validate the information and answer further questions to clarify and to add to the information already in the case study.

The fourth pillar organizations considered in this study comply with the previously stated definition of Liljemark (2004). Table 6.2 summarizes the requisites that the four organizations chosen had to accomplish in order to be considered a fourth pillar organization. Apart from definition, the criteria for inclusion were that the organizations had been operating for at least 4 years and could be considered successful. Success was assumed if they had been identified as good practices within the CLIQ project.

5 Results from Case Studies

The main description of the organizations analyzed is presented in Table 6.3. They belong to three different countries, namely, Spain, Holland, and Sweden. Two of them serve industries in the secondary sector, such as steel and materials in general,

Table 6.3 Description of the fourth pillar organizations in the study

	Centre Tecnològic Manresa (CTM)	Centre of excellence for sustainable water technology (WETSUS)	Future Position X (FPX)	Triple Steelix industrial region
Country	Spain	Holland	Sweden	Sweden
Description	CTM carries out research, development, and technologic innovation projects and provides specialized services regarding analysis, innovation support, and industrial training	WETSUS is a facilitating intermediary for trend-setting know-how development in the water treatment technology	FPX is an independent society that supports and develops the member companies' competitive abilities and presence in the market within the Geographical Information Systems (GIS) field	Triple Steelix is a regional developmental effort to further strengthen the successful steel industry in Bergslagen
Year of creation	2000	2003	2004	2004
Sector	Materials technology, environmental technology, and innovation support	Water technology	GIS	Sheet steel, stainless steel, machining, services, and subcontracting
Legal form	Foundation	Foundation	Cluster organization	Cluster organization
Number of workers in 2008	83	Not available	12,000 (including workers of companies)	5 permanent workers in the organization and 16 engaged collaborators

Table 6.4 Importance of partners in the fourth pillar organizations

	Industry	Academia	Government	Financial institutions
CTM	+	+	++	++
WETSUS	++	++	+	+
FPX	+	+	++	+
Triple Steelix	++	++	+	+

+ Indicates average importance of the type of partner within the fourth pillar organization, ++ Indicates stronger importance of the type of partner within the fourth pillar organization

one deals with water processes, and the other with software. They had been operating for 4–9 years when they were studied, which means that they were relatively young organizations but with an already recognized trajectory in their correspondent innovation systems. The Swedish organizations have the legal form of cluster organizations, and the other two are foundations.

As regards to the partners participating in the fourth pillar organizations, their role had a varying importance for the different fourth pillars as presented in Table 6.4. Industry and academia were strong in the case of WETSUS and Triple Steelix while government was stronger for the other two.

From the profiles presented in Table 6.4, we can infer two typologics. Thc first one includes Triple Steelix and WETSUS. These fourth pillars are different from the others in a number of ways. Government and financial institutions play a less important role in the fourth pillar because the industry and the market mechanisms are stronger in their sectors (water industry and steel). The second model is based on the case of CTM and FPX, where the role of the public sector has been stronger. In both cases, the foundation of the organization is due to public leadership, providing funds in the case of CTM and also creating public expertise in the case of FPX, when they establish the Swedish mapping, cadastral and land registration authority in the city of Gävle, home of FPX. The strength of the public sector is paired with the lack of strength of its original industry and academia (Fig. 6.2).

We should note that although FPX and Triple Steelix belong to the same country, they are classified in two different models. We attribute this difference to the need of fourth pillar organizations to be adapted to the characteristics of the preexistent innovation system. It is important to mention that the initial status of the triple helix participants is then effectively changed by public action. In the case of Model 2, public leadership in creating the fourth pillar organization has proved extremely successful, contributing to building a much stronger academia and industry in its respective fields of activity.

A deeper analysis of the role of the fourth pillar for each of the two models uncovers that in the Model 1 (Triple Steelix and WETSUS) universities do a lot of R&D because the academic network of the fourth pillars is very strong, and the industries they work with need a research infrastructure, such as labs, that is more easily provided by universities. This motivates a high level of subcontracting with universities that can be channeled via the fourth pillar. Figure 6.3, which is based on Rogers (2003), presents the role of the fourth pillar within the innovation process, mostly coordinating research and innovation (development, production, and launching) but leaving ideation and marketing to firms and R&D to universities.

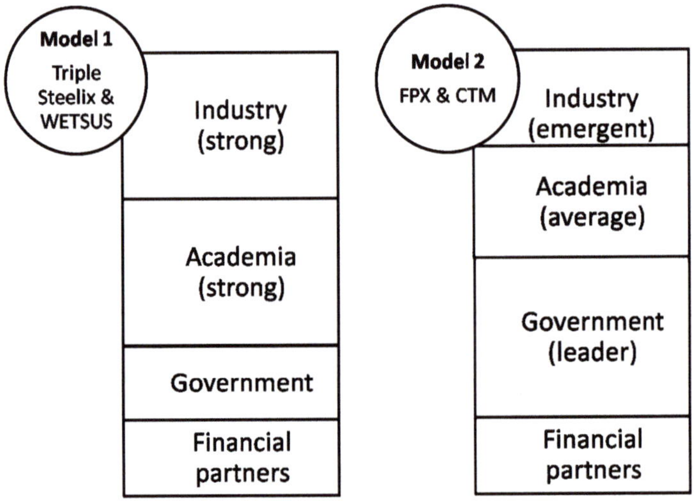

Fig. 6.2 Fourth pillar organizations' models

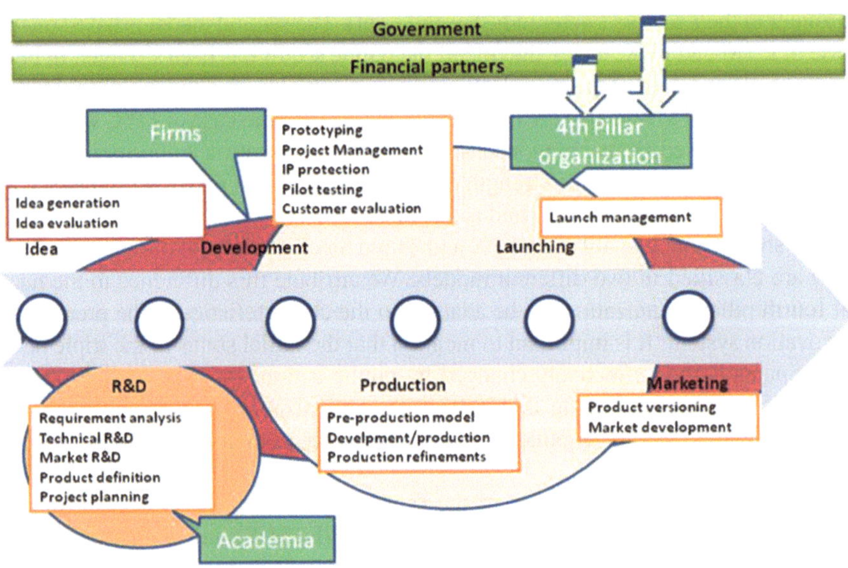

Fig. 6.3 Model 1 of fourth pillar organizations innovation flow: Triple Steelix and WETSUS

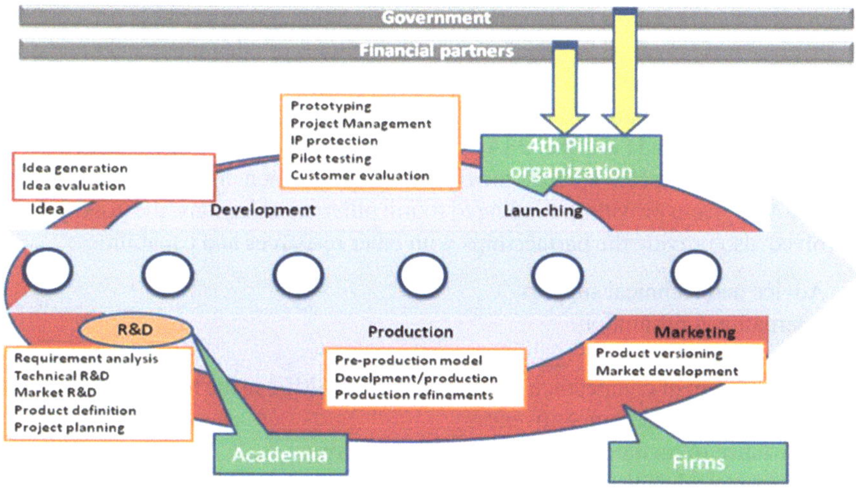

Fig. 6.4 Model 2 of fourth pillar organizations innovation flow: FPX and CTM

Although in Model 1 the sectors were relatively large and competitive, Model 2 is found in emerging, nonconsolidated sectors, which have weaker market structures, since industries were relatively new (i.e., geographical information systems, material technologies) compared with the others. For this reason, the role of the government and financing institutions is substitutive of the market structures and has a vital role in supporting and funding R&D projects in fourth pillar organizations. In this case, the fourth pillar has a more important role than just coordinating the research; it has to act as a stimulator for this research and technology transfer. Because the academic network of these organizations is also weaker, the fourth pillar takes the lead in the research programs. This means that the role of fourth pillars goes a step further and integrates backward with the typical research activities that a university carries on in Model 1. This is facilitated by the fact that the infrastructure needed to do research in their technological sectors can be more easily provided by the fourth pillar. Both CTM and FPX do R&D in-house while the other fourth pillars basically subcontract these activities to their university partners. This second model is represented in Fig. 6.4, which is based on the work by Rogers (2003).

6 A European Model for Collaborative R&D Projects

When analyzing the four studies, it became evident that there were some common differences across cases in respect to the existing analyses of fourth pillar organizations. According to Johnson (2008), government plays a minor role in the innovation process, especially regarding its role in triple helix partnerships. In Fig. 6.1 of this paper, a particular structure for supporting triple helix collaborative R&D has been presented based on Johnson (2008). In this figure, government is presented as "other partners" together with other research institutes or other companies and does not serve an important role in the triple helix representation. Johnson's study is centered in Canada, and fourth pillar organizations have mainly been studied in countries, like Canada and the USA.

In our European case studies, the role of the government appears to be very different. Apart from providing funding to fourth pillar organizations, the governments involved also provide the partnerships with other resources and capabilities:

- Advice and technical support
- International promotion
- Research contacts
- Development of competence in order to support SMEs in the region and to assure their international competitiveness
- Assistance in revitalizing the image of the regions
- Increased exchange of information between university, SMEs, and large companies utilizing the strong existing base of knowledge in the regions
- Strengthening the ability of innovation through the development of new networks and meeting points
- Entrepreneurship stimulation

The case studies provided some lessons about the role of the government in fourth pillar organizations:

- Public innovation policy plays a vital role in S&T by catalyzing and feeding the system with money.
- The authorities do not rely only on the invisible hand of the market. National, regional, and local governments play a crucial, irreplaceable role in stimulating innovation in all countries, where the fourth pillar organizations studied come from.
- The role of the government is limited gradually when the other two actors (industry and academia) assume larger roles in the S&T sector.

The model also draws out another major difference from Johnson's model, which is the role played by financial partners. The European model proposed places more relevance on financial institutions (especially banks) when supporting fourth pillar organizations. In our model, the financial institutions provide a great part of the funding needed for developing the research projects of the triple helix partnerships. They provide other financial services, like loans, insurances, etc., to facilitate the project's development.

Financial institutions provide service as intermediaries of the capital and debt markets and their role is more relevant in the EU countries, where the market mechanisms are not as perfect as those in the USA or Canada. Because the market is not so strong in the EU countries, other mechanisms to help and provide funding for innovation activities need to be introduced. The European model for triple helix collaborative R&D projects is, therefore, modified by the major role played by governments and financial institutions.

This justifies the modification of the existing modulation of fourth pillar organizations. We propose a different model for collaborative R&D partnerships, modifying

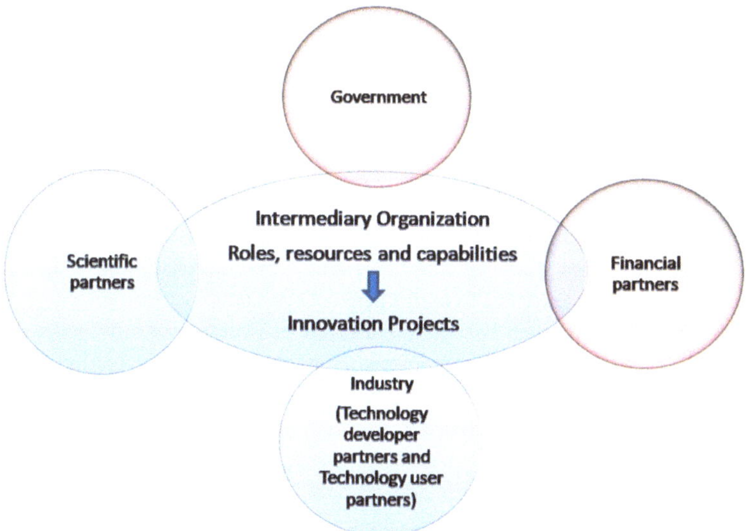

Fig. 6.5 A European model of fourth pillar organizations for supporting triple helix collaborative R&D projects

the proposal of Johnson (2008) to better suit the reality found in our four cases. Figure 6.5 depicts the modified model, including the role of government.

7 Conclusions

This chapter had the motivation to study how fourth pillar organizations could help public policy makers to improve the interaction among the different actors in the innovation process and to increase the implementation of successful innovations. We have started from the basic assumption that triple helix partnerships could be a valuable strategy in helping this collaboration by achieving the primary goal of managing innovation and successfully commercializing new technologies.

This chapter has implications for national, regional, and local governments as it provides exemplar cases on the role that public administration should have when promoting innovation in triple helix partnerships. The creation of fourth pillar organizations and other collaborative tools of innovation support needs to be included in the innovation agendas of these public institutions. National and European institutions can also draw information on how to promote, fund, and get involved in innovation partnerships. As noted in the current Europe 2020 strategy, the main interest of governments should be in increasing their relevance with other partners in order to build a cohesive network able to efficiently develop innovations. Policy deployment is expected to follow this direction.

When looking at the fourth pillar organizations that we have studied, the role of the government appears to be very different from other studies done on these organizations. Apart from providing funds to fourth pillar organizations, the governments involved also provide the partnerships with advice, technical support, international promotion, research contacts, and entrepreneurship stimulation. For this reason, we point to the existence of a European model for collaborative R&D partnerships, which better suits the national realities regarding innovation issues in the EU and that emphasizes the role of government in innovation. The model builds on the proposal made by Johnson's (2008) model of fourth pillar organization but is different in a number of ways. Johnson's model was designed to suit countries, like the USA or Canada, with strong market structures. Our model applies to European countries, where the market is not so strong and the role of governments and financial institutions has proved to be essential to facilitate R&D projects.

Moreover, we distinguish two different types of European fourth pillars created to complement the preexistent characteristics of the triple helix environment. The first model is adopted for a strong, competitive, and consolidated industry, where the role of the fourth pillar organization is based on the coordination of the activities developed. The second model is found for emerging technological sectors, where the fourth pillar goes a step further and acts not only as a coordinator but also as a stimulator and R&D infrastructure provider.

The four European cases analyzed in this chapter are meant to be illustrative and exemplar for policy makers to better promote and manage innovation among triple helix actors and for business, clusters, and industry in general, to extend and adapt these findings to their particular industrial context.

References

Audretsch, D., Grimm, H. and Schuetze, S. (2009). Local Strategies within a European Policy Framework. European Planning Studies, 17, 3.

CANARIE Incorporated, CMC (Canadian Microelectronics Corporation) and Precarn Incorporated, (2003). Accelerating Canadian Productivity Growth: Stimulating Innovation through 4th Pillar Organizations. Brief to House of Commons Standing Committee on Finance.

Carayannis, E.G. and D.F.J. Campbell (2009). 'Mode 3' and 'Quadruple Helix': toward a 21st century fractal innovation ecosystem. *International Journal of Technology Management*, Vol. 46, Nos. 3/4: pp. 201–234.

Cooke, P., Heidenreich, M. and Brazcyk, H.J. (Eds) (2004). Regional Innovation Systems: The Role of Governance in a Globalized World, London: UCL Press.

Dalziel, Margaret, fourth pillar Organizations in Canada (September 15, 2005). Available at SSRN: http://ssrn.com/abstract=1518863.

European Commission (2009). Policy Highlights Overview, available at: http://ec.europa.eu/enterprise/policies/innovation/index_en.htm.

Garofoli, G. and Musyck, B. (2001). Innovation policies for SMEs in Europe: Towards an interactive model? Regional Studies, 35, 9, 869–872.

Garofoli, G. and Musyck, B. (2003). Innovation policies for SMEs. An overview of policy instruments, in: B. Asheim, A. Isaaksen, C. Nauwelaers & F. Todtling (Eds). Regional Innovation Policy for Small-Medium Enterprises, Cheltenham: Edward Elgar.

Holbrook, A. (2002). The Federal Government Role in the National System of Innovation, available at: http://www.innovation.cc/salon/2002/2002-09-06-holbrook-slides.htm.

Johnson, W.H.A. (2008). Roles, resources and benefits of intermediate organizations supporting Triple helix collaborative R&D: The case of Precarn. Technovation, 28, 495–505.
Johnson, W.H.A. (2009). Intermediates in Triple helix collaboration: the roles of fourth pillar organizations in public to private technology transfer. Int. J. Technology Transfer and Commercialisation, 8, 2,142–158.
Liljemark, T. (2004). Innovation Policy in Canada. Strategy and Realities. Swedish Institute for Growth Policy Studies.
MacGregor, S.P, Marques-Gou, P. and Simon-Villar, A. (2010) Gauging Readiness for the Quadruple Helix: A Study of 16 European Organizations. Journal of the Knowledge Economy 1:173–190.
Musyck, B. and Reid, A. (2007). Innovation and regional development, do European structural funds make a difference? European Planning Studies, 15, 7, 962–983.
Rogers, E. M. (2003). Diffusion of innovations (5th ed.). New York: Free Press.
Stark, M. and Wolf, C. (2007). From idea to innovation. Success factors and challenges in the innovation process. Fraudenberg New Technologies, available at: http://www.fnt-kg.com/pdf/brochures/FNT_03_2007_en.pdf.
Yin, R.K. (1989). Case Study Research: Design and Methods, Sage Publications, Newbury Park, CA, 166.

Chapter 7
Strategic Process of Change: A Multiple Network Game—The Rohner Textil Case

Nigel Roome and Céline Louche

Abstract Few companies have successfully undertaken wholesale change to embed corporate responsibility for sustainable development into their organization and business model. Existing cases suggest this change involves companies in complex processes of organizational and social learning, innovation, and change that play out within a range of different networks of relationships. This chapter examines Rohner Textil, a company that embarked on strategic change toward a more sustainable approach. The organizational and social processes and steps the company went through are described. The case highlights the critical interaction between the company and other actors through three different networks: an industrial network and a knowledge network, which provided new concepts for inspiration, and an internal network of ideas and actions, which would help define and shape change. We argue that the success of the company was dependent on the ability of management to effect organizational leadership in and through these three types of networks. The case provides evidence of what that involved and points to the complexity of the tasks performed in these action-learning networks. It is argued that companies working for change toward sustainable development can be better understood through an action-learning network model of innovation and change.

N. Roome (✉)
Vlerick Leuven Gent Management School, Vlamingenstraat 83, Leuven 3000, Belgium
e-mail: nigel.roome@vlerick.com

C. Louche
Vlerick Leuven Gent Management School, Reep 1, Leuven, Belgium
e-mail: céline.louche@vlerick.com

S.P. MacGregor and T. Carleton (eds.), *Sustaining Innovation: Collaboration Models for a Complex World*, Innovation, Technology, and Knowledge Management,
DOI 10.1007/978-1-4614-2077-4_7,

1 Introduction

Sustainable development was originally described as a frame of change (Brundtland Commission 1987). This emphasis on change through collaboration was reinforced in the universal declaration of the principles of sustainable development: Agenda 21 (Agenda 21 1992). The significance of sustainable development as a process of organizational and social change was explicitly recognized by the first commercial organization to design a strategy for sustainable development in the early 1990s (Roome and Bergin 2006). These processes were identified as a key element in the greening of business (Gladwin 1993); however, very few examples in the past 20 years describe how these processes occur, and how they link the economic and environmental aspects of business and contribute to sustainable development. Early commentators suggest they involve technological innovation (Roome and Winn 1993) and collaborative learning providing for innovation that operates through networks that span business organizations and other actors (Roome and Clarke 1995). This is understood as a complex and difficult process (Hall and Vredenburg 2003; Roome and Bergin 2006). The implication is that these learning, innovation, and change processes not only occur at the company level but also require companies to engage in fundamental change in their relationships with a wide range of other actors through networks (Clarke and Roome 1999; Roome and Bergin 1998).

Understanding these processes is particularly important at the present time as attention is now being focused by the European Commission on the competitiveness of business, as well as its contribution to sustainable development. Innovation and change provide the means to combine these ideas. This link was recognized in the Lisbon Strategy, but the strategy and its goals failed for structural reasons (Kok 2004). An alternative explanation is that the processes and the skills (or dynamic capabilities) that underpin these goals are not well understood. They are complex, uncommon, and too rarely subject to research. As a result, the know-how on which those processes of change are based is not part of conventional management education and training (Roome and Cahill 2001). This suggests the problem is not only structural but also concerned with a lack of know-how. That is consistent with the notion that sustainable development represents a new paradigm that has mostly been interpreted through the lens of the existing paradigms of business. That view is supported by the way that sustainable development and innovation was understood in the Lisbon Strategy in terms of the environmental technology industries and the development of the concept of win–win strategies (Kok 2004).

This chapter takes as its starting point the case of a company that embarked on a strategic process of change, namely, Rohner Textile. The company promoted a more sustainable approach to its business model using rather different concepts. There are very few examples of companies that have successfully undertaken wholesale change that has led to the embedding of corporate responsibility (CR) for sustainable development into the organization and its strategy. Interface is a well-known example of such a company, but it only began to develop its approach in 1994 (Interface 2011).

Rohner has attracted quite some interest in the literature. McDonough and Braungart (2002) refer to Rohner as an illustration of their Cradle to Cradle concept. The CEO describes how Rohner addresses sustainable solutions in product development (Kälin 2001). The company has also been the object of several teaching cases (Gorman et al. 1997, 1998; York 2006; Khan and Steger 1999). The cases and articles address a wide range of issues related to ethics, technological solutions, innovation, decision making, costs and risks, environmental impacts, strategy, and competitiveness. However, none of this earlier work has addressed the idea that sustainable development is a social or multiactor process that business can only contribute to by working on innovation and change with other actors through highly networked relationships.

This chapter aims at better understanding the complex process of innovation and change with its emphasis on the organization and its interaction with other actors. It especially focuses on the different networks of relationships that are worked through to effect sustainable development. Based on the evidence of the case, we argue that change toward sustainability requires an orchestration by a focal company of multiple, dynamic, and intertwined networks. It suggests this is founded on a sequence of nonroutine activities, and it is supported by some key practices and beliefs that provide the dynamic capabilities on which the business contribution to sustainable development is based.

2 Method

This study is based on an in-depth single case study (Yin 2003) to explore the dynamic and interdependence between types of networks the company had to influence in order to effect change. Siggelkow (2007) claimed that a single case study can contribute to existing knowledge through the deepening or widening of current understanding. Case study research is regarded as appropriate for the investigation of questions of how and why, especially when the concern is to study phenomena over which the researcher has little or no control (Yin 2003). Case study research is particularly appropriate when the boundary between the phenomenon under investigation and its organizational and social context is blurred, which was the case in this study. This is an open-ended exploratory study aiming at examining how and why change leadership for sustainable development through CR happened at Rohner and more especially to identify the role that networks played in the dynamic process of change. The data used are qualitative.

2.1 Case Selection

Rohner Textil AG is a leading Swiss company in high-quality upholstery textiles mainly for commercial applications. Rohner was chosen as case study because of its pioneering role in integrating CR into its business model and business practices, making the company rather unique. Rohner managed to balance ecology and economy to an unprecedented level, making significant effort to include social issues in this balance. In particular, their efforts in building redesign, together with the development and design of manufacturing processes and the product itself, are aimed at the development of leading-edge sustainable products.

In a way, Rohner can be classified as an "extreme case" as defined by Pettigrew (1990). Extreme cases facilitate theory building because, by being unusual, they can illuminate both the unusual and the typical (Patton 2002; Eisenhardt 1989). Pettigrew argues that in extreme cases the dynamics tend to be more visible than they might be in other contexts. Rohner presents characteristics of an extreme case, which provided us with the opportunity to study the process of change arising from a strategic approach to CR. The company is an extreme example not only because it began its change process earlier, it also changed its own business model, and it had a significant role in changing the approach to sustainability and business of other actors through collaboration and learning.

2.2 Data Collection

The case analysis is based on primary and secondary data. It was originally written in 2004 and updated for the purpose of this chapter. Multiple data collection methods were used for the case allowing triangulation (Eisenhardt 1989). Company visits (approximately 4 days in total) were conducted that served three purposes: (1) field observation, (2) interviews with key senior personnel, and (3) collection of written material. In total, six people were interviewed in February 2004 including the CEO and senior managers dealing with strategic and operational issues (quality, environment, research and development production, dyeing, and maintenance). An additional interview was conducted with Rohner's former textile designer in 2011 to obtain an update of the latest developments in the company. Interviews were recorded and transcribed. Secondary data included the company's reports, Web site, and other publicly available information.

The case drafted in 2004 was sent to the company contact for verification of details in the case. At this time the contact was invited to answer questions about the key factors that contributed to the success of the company in CR, the main obstacles that had to be overcome, and the key learning points for the company to emerge from the process of managing CR.

2.3 Data Analysis

The data collected was organized and prepared following a case research protocol, which aimed at exploring how the company was learning to interact and respond to its changing context in a way that was new to the managers. This started with the recognition of the need for change through the creation of a shared vision of the future of the company, which was supported by the development of new concepts that were brought into the company and then combined with ideas from inside the organization. The commitment to the vision and the new concept was supported through ideas from the workforce that would move the company as a whole toward the chosen vision. This affected other actors beyond the company. These actors included a range of first-tier suppliers to the company, such as chemicals firms that provide materials for dyes and wool farmers. It involved companies that designed and installed machinery and facilities, like the dye house, weaving loom or water system in the company's factory. Rohner's actions affected the company's customers, local environmental regulators, and its neighbors. The modifications to the production process and products reduced the risks of pollution to many users of water carried in the Rhine, as the company was located in the watershed at the source of the Rhine. However, Rohner was also part of a knowledge network that brought ideas to the company about how to create change and Rohner became an example and vehicle for this knowledge to be taken to others. Finally, as the company added know-how about sustainable-design of fabrics, it ended selling that knowledge through license agreements to other producers of fabrics as well as producing its own fabrics.

Given the exploratory nature of the case study, we did not apply any specific analytical techniques to the material but rather let the findings emerge from the data themselves. All data gathered were compiled into one single document describing the company's historical and strategic path related to CR including the drivers for CR, the strategic decision making processes used in the company, and the day-to-day practices of CR.

3 The Process of Change: Reconciling CR and Competitiveness

At the time the case was originally written, in 2004, Rohner Textil AG was a very successful manufacturer of high-end upholstery fabrics. Based in Balgach (St. Gallen, Switzerland), the company was founded in 1947. Rohner was internationally recognized for its qualitatively outstanding products—especially for Climatex® Lifecycle™ and Climatex® LifeguardFR™, biologically recyclable, environmentally sound fabrics—and its commitment to sustainable development. Between 1994 and 2004, the company won 20 prizes and awards at the national and international level for its environmental achievements.

In 2004, Rohner had 30 employees and produced around 250,000 meters of upholstery fabric. In 1988, Rohner was taken over by the family owned Foster Group, an enterprise of 700 employees and consisting of five European textile mills with specialities that ranged from upholstery fabrics to men's socks and sports socks to the main business in embroidered products. And in 1999, Rohner Textile AG was taken over by the Swiss company Lantal Textiles. Despite those major ownership changes, Rohner managed to maintain its identity and independence.

Rohner's success was the result of a long process of change. This section describes the development of Rohner between 1981 and 2004 and provides an update for the period of 2005–2011. Five main periods are used to frame the 23 years. During the first period, Rohner had to face tough challenges that were both economic and environmental. From an economic viewpoint, Rohner faced fierce competition from non-European textile manufactures. From an environmental viewpoint, it faced a number of concerns including increasingly stringent regulations over water use and pollution from dyestuffs, as well as concerns over vibration effects on local neighbors from the use of its machinery. These problems seriously threatened the survival of its business. After a first step of "cleanup" but also as a result of thorough reflection on the environmental and economic pressures on the business, Rohner decided to go one step further in their shift toward high-quality products which would also show environmental responsibility around 1990. Rohner then began an ambitious and difficult journey to pursue high-quality, environmental intelligent products that would involve significant product and technological innovation and change. Around 1995, the innovative approach of the company started to be rewarded as a brand new product was launched following strict environmental criteria. The third step was to implement the same process for all products of the company. In the late 1990s, as Rohner was starting to collect the fruits of its efforts, the company entered the fourth step, as the managing director continued to push the boundaries further toward sustainability to integrate new dimensions. However, during recent years, the textile market changed and brought new challenges to Rohner. The following paragraphs describe the five periods.

3.1 1981–1992: Challenging Systems for Survival

Albin Kälin, Rohner's managing director, joined the company in 1981 and took over as managing director in 1989. During this period the company had to face two main pressures that forced it to move toward sustainability: a threatening economic situation and a number of constraints related to the local context.

After 1980 the European textile industry, especially in Germany, Switzerland, and the UK, suffered significant decline. The traditional mills in these countries faced intense price competition from new mills in India and the Far East, which were able to combine low labor cost with the quality available from investment in new technology. Many European textile companies closed their businesses due to severe price competition. In addition, European textile firms faced increasingly stringent environmental regulations and costs to reduce their environmental impacts. In response, many textile companies outsourced their production to low-wage countries.

Another set of pressures came from the location of the company. Rohner was located in a building constructed in 1911 and was under a series of building conservation orders. The company was not allowed to make changes to the architectural fabric of the buildings without prior consent. In addition, the company was located in the water-catchment at the source of the river Rhine near Lake of Constance, a major water-drinking reservoir for local people and the beginning of a river system that spans six European countries and more than 1,200 km. Moreover the building was located in a mixed residential and small business district. Consequently, the firm was subject to strict noise restrictions and high water protection standards. Those restrictions hampered the company to improve productivity, as it would have meant more noise and vibrations. Relocating the firm was not an economically feasible option. In addition, any accident in this protected location and surrounding could seriously damage the company's reputation and future license to operate.

Those challenges made clear to the company that a gap between societal and environmental expectations and the company's situation and practices was widening. Something had to change, or it would not be able to survive. Rohner managed to provide creative responses to those challenges and turn them into opportunities to operate and produce textiles in radically new ways. Albin Kälin's rebellious personality played a crucial role in pushing the boundaries further and breaking the mold of established thinking.

The first challenge was answered by embracing a strategy of high-quality specialized products, aesthetically valued and efficiently produced upholstery fabrics. This direction was taken in the very early 1980s and paid off as Rohner emerged as a worldwide leader in this area.

With regard to its locational challenges, Rohner managed to exploit the potential of the space of its historic building to the full through technological innovations. Kälin built a new dye-house that fitted the needs for capacity and flexibility and drastically reduced noise, vibration, water, and energy consumption by 40%, despite the skepticism of the larger manufacturers of dyeing equipment. One of the most radical steps in the technological changes was the elimination of cotton from the product line at the end of the 1980s. This was a necessary step to reduce the company's environmental impacts. As a result of those actions, the company significantly decreased its water and waste-water costs.

During this period, the company addressed a wide range of economic and environmental concerns, but beyond that, the search was for mainstream solutions found within the industry. Updating machinery was not sufficient for Albin Kälin and especially would not guarantee that his product quality would improve over time. It was also important to build a company culture that would make reducing the company's environmental impact everybody's goal and develop a strategy around that idea. This became one of Albin Kälin's priorities when he became managing director in 1989.

3.2 1992–1995: Exploring New Concepts

In 1992, the company made a commitment to this new path: It shifted from a reactive environmental management approach based on incremental change to a proactive stance requiring a thorough and radical redesign of the company's processes and products. This would move the company away from solutions that it could control by asking other companies to provide infrastructure and technologies through investment by Rohner toward working through more open and collaborative mechanism involved in co-designed change. The first step involved subjecting all of Rohner's products to tests in order to obtain the Eco-label Öko-Tex Standard 100. Rohner's products passed these tests. However, all the environmental issues related to dyeing and other manufacturing processes used in manufacture were not resolved despite the fact the products met the eco-label standard. It was found that some environmentally unacceptable chemicals remained in the finished products. Although a closed-loop production process would solve the concern over contamination to the local area, this only meant that its customers would have some possible contamination in the products they bought. This led Rohner Textil to develop a strategy designed to use ecological principles as a basis for their business and product development.

In order to formulate and develop this concept, Kälin took half his staff to a restaurant in the hills near Balgach. This restaurant had no electricity, no gas, and no running water. The restaurant was lit with candles, the cooks worked with a wood stove, and water only dripped from the tap. In this exceptional situation, Kälin raised some questions about how to work effectively in a specific situation of scarce and limited resources. The brainstorming session led to many ideas that were formalized in the first written strategic concept: Öko–Öko 1993–2000. At the heart of this approach was the need to invest in ecological efforts while simultaneously investing in new equipment as a basis for productivity gains. It was recognized that over 10% of Rohner's turnover had to be invested in new equipment to increase productivity and maintain or improve competitiveness.

Around this period a breakthrough was discovered when Rohner's tax consultant identified that the region of St. Gallen provided a special tax regime for the depreciation of capital related to environmental investments. This tax incentive provided Rohner's management with the possibility to justify environmental investments that the board of directors could agree upon. Investments had to be planned ahead of time. Beginning in 1993, Rohner began to identify environmental investments in its annual budgets. The investment amounted to 1% of the total budget. In addition, environmental costs were included in calculating the cost of producing the product. The established accounting systems in the company had to be reformulated to incorporate these measures. The first environmental investments were implemented in 1995.

At that same time in 1992, Susan Lyons, design director at DesignTex, Inc., a subsidiary of Steelcase, Inc., located in New York, approached Rohner Textil. Lyons had been studying innovative suppliers and learned about the unique strategic initiatives at Rohner Textil toward environmentally optimized production. Mrs. Lyons proposed that a DesignTex team work with Rohner to develop an "environmentally intelligent" line of textiles, focusing on the product itself, not just the production process. In order to provide the necessary conceptual and technical knowledge for this approach, she suggested that Rohner collaborate with DesignTex, run by the American William McDonough and his friend and partner Michael Braungart, a chemical engineer and founder of the German Greenpeace Chemical Division.

Their company's goal was to bring about a vision and design for a new industrial revolution: the clean revolution. When William McDonough first visited Rohner in 1993, he spoke about the idea of "waste equals food." For him, that meant no process or product should be produced that was harmful or contaminated. Kälin embraced this concept as part of the approach needed to shape the content of the emerging vision for Rohner.

Together they engaged in a new and innovative route toward sustainable fabric design, seeing how the concept of "waste equals food" would relate to the products Rohner produced or might produce and the way they were produced and used. But the route was not easy. To do this, every aspect of the company's fabric's development had to be evaluated, and every chemical and material input had to be identified and inspected. For this purpose, Rohner's dye suppliers had to cooperate in identifying the chemical composition of the materials they supplied. This turned out to be impossible. Sixty chemical companies worldwide were contacted, none of which were willing to share the information on the chemical processes to create their products used by Rohner—except Ciba-Geigy. From the information submitted on about 1,600 dye chemicals, Braungart and the Environmental Protection Encouragement Agency (EPEA), an independent environmental institute in Hamburg, selected 16 dye chemicals, which met their strict design criteria and addressed effects on human health and the environment. All colors, with the exception of black, could be developed from the 16 selected dyes.

Eighteen months later in 1995, the product Climatex® Lifecycle™ reached the market. During the process, a complete technical reorientation of the product was implemented with the assistance of Michael Braungart and the EPEA. EPEA was given the possibility to engage the production network and talk with the suppliers about the quality of their supplies and eventually to recommend modifications. The first collection called "The William McDonough Collection" was released in the USA with an informational booklet entitled "Environmentally Intelligent Textiles." The product was launched in Europe in 1996.

In contrast to the first phase from 1981 to 1992, this short period of 3 year up to 1995 involved the company in intense face-to-face collaboration to inject new ecological concepts that would provide the ground for future development of the company and its products. The ambition also shifted from process improvement to improvements in the products themselves.

3.3 1995–1999: Formalization and Diffusion of the Concept

In 1995–1996, clear structures were developed within Rohner that combined ecological considerations with business concerns. Rohner was one of the first companies to introduce an environmental management system that satisfied both ISO 14001 and EMAS 1836/93. It was the first Swiss company to establish an environmental management system according to the EMAS, and in 1994, Rohner received ISO 9001 certification.

Although these management systems were in place, the company realized that, in order to make solutions really effective, it was important to understand its interaction with all the systems with which it was connected. The management systems were necessary but not sufficient for design for sustainable development. A key aspect of this design approach was to know more precisely what the company's products contained and the materials they were made from. Rohner developed an eco-controlling concept: a quantitative assessment of the environmental impact of each of its products. This provided the basis on which to found an ecological product development (design) process for their entire product range and enabled the R&D department to focus better on environmental improvements for quality products, or as it became known, sustainable product redesign.

In 1997, 3 years earlier than planned, Rohner managed to reach the goals it had set in the plan for 2000. Kälin took this opportunity to move even further in his sustainability vision supported by his new concepts, especially to integrate social concerns into his strategic orientation. Again, he took half his staff for a 2-day seminar at an abbey in the mountains of Austria. The surroundings were intended to provide for an atmosphere of solidarity. And again, Kälin raised questions about sustainability. During the session, some very practical, concrete ideas came up, such as the idea of closeness, openness to criticism, and wages issues. The seminar was very emotional; everyone really opened up. All inputs at the meeting were recorded and documented. This retreat provided the basis for the second strategic concept: *The path towards a sustainable company, Rohner Textil AG 1998–2008.*

Although company profits were good and the supervisory board approved the second strategic concept relatively easily, some issues still needed clarification. In the seminar, it emerged that the executive structure needed to become less hierarchical, including external consultants (to include the EPEA, ISO-auditors, accountant, etc.). The idea of including external consultants in the executive structure gave the consultants a position of trust, as well as belonging and interdependency (closeness). All the consultants were honored and proud. However, this structure was very unconventional in relation to normal business structures and was difficult for the supervisory board to accept.

In 1998, Rohner also formulated its credo and developed an Index of Sustainability. The index served as a central information tool to analyze the environmental soundness of Rohner's products evaluating the toxicological and ecotoxicological impacts of products. It included aspects of resource renewal encompassing all material flows at all process levels in the mill and dye house but also along the production chain. The index was made available to product users and customers and other interested parties in order to give them a better understanding of the progress in the design of a product toward sustainability.

This period was characterized by collaboration intended to elaborate and make operational the new concepts laid down in the period before. This phase of organizational development involved intense face-to-face collaboration if the new concepts were to be adapted to Rohner's situation, and understood by management and employees. Moreover, open communication would also be needed if employees were to contribute ideas on how they could change their routines and practice in ways that would translate the adapted ecological concepts into realities for Rohner.

The period also signaled the need to push suppliers to support the move toward new products. This too required collaboration with Rohner seeking change in the larger companies that supplied it.

3.4 1999–2004: Pushing the Limits Even Further

Rohner succeeded in developing a new generation of Climatex® Lifecycle™ just in time for the new millennium. This new product provided flame retardant capability in addition to the fabric's reliable environmental and aesthetic features. The product is named Climatex® LifeguardFR™. It was designed for applications in public spaces and for worldwide air, train and water transportation applications.

The direction of Rohner's product development foreshadowed tremendous market opportunities in a range of different market segments. An increasing number of office furniture and transport service supply companies made commitments to environmental aspects in their products and services. This created a customer base eager for a good price/quality specification and who wanted added value in terms of environmental claims and specification. However, the opportunity for the company to grow its production facilities on site to match customer demands was limited because of the company's location.

Nonetheless, Rohner identified other assets that might permit revenue growth. It had developed a key understanding of its products, their material components, and profile. It had positively defined its supply chains in terms of their contribution to the sustainability of the final product, and it had developed a committed customer base behind Climatex® Lifecycle™ and Climatex® LifeguardFR™. Rohner had tacit knowledge that was hard to copy, technical knowledge, managerial ingenuity to make consistently high-quality products, a growing reputation, and worldwide brand recognition. In developing its new product, Rohner Textil had built a "know-how platform." This led to recognition of the value to the company of knowing how to "design for sustainability." Design for sustainability created value—the need then was to find a business model that would capture some of that value for the company.

Despite the complexity of their supply chain, Rohner saw that their knowledge was transferable in the form of license agreements with other textile companies. Licenses included both production and marketing and enabled the company to move ahead in terms of revenue growth and maintain its market differentiation. Rohner, as a small company, planned to secure global market leadership in different business segments while retaining the market segment for residential home interiors and contract furniture from its production at Balgach. The licensing process provided a means for Rohner to exercise environmental and product quality control ensuring that products conform to expected standards in different production contexts. EPEA monitors the process from the scientific point of view.

In 2003, Rohner was confronted with a market downturn and a set of financial concerns. Nevertheless, Kälin continued pushing the boundaries further and reinventing textiles. He started to work on a revolutionary new strategic concept, which was initiated some 8 years before. Thinking in product cycles, Rohner had managed to take control over a substantial part of their product cycle: from cradle to user and back. In extending the closed-loop concept, Rohner did not just sell textile fabrics that did not leak resources to their neighborhood, their customers, or the users of their customers' products, but they sold textiles with ecological integrity and functionality. The next step was to consider the logistics required to close the product/consumption cycle completely so that the company managed the flow of materials throughout the production/consumption system. This involved product take-back from their customers (office equipment makers and others) and their customers' customers (companies that buy/lease office equipment). However, there were two major problems. The first concern was to make sure that used materials were not contaminated through use and second, international regulation prohibited the return of industrial products in biological cycles of this kind. International regulation was designed for thinking that was based on the throughput of materials. However, when Kälin was giving a presentation to an MBA class, he met an expert on waste issues. Together, including EPEA, they wrote a project proposal to tackle both problems. The plan proposed a partial recovery of materials when used products were taken back. This recovered the energy locked in the material while the material residue became clean fertilizer. Curiously this way forward provided products (energy and fertilizer) that were more valuable than the biomass of the waste material. In the case of the issue of international regulation, Kälin said, "We have advanced plans to challenge legislation."

In late 2003, Kälin was invited by New Zealand's sheep farmers to talk about his approach as the company's products contained the finest New Zealand wool. They asked him to tell them about his closed-loop concept. The farmers were so interested in the model that they founded a Climatex Club that started working on the idea of sustainability. In this way, the concept of closed material loops that had been applied in Rohner's textile products originating in Switzerland was beginning to affect agricultural practices in New Zealand. In the same way, Rohner challenged the chemical industry not only about the ingredients of the dyestuffs it supplied, but the chemical industry was now being pushed or inspired to develop new pesticides, medicines, and fertilizers for use in agriculture. Said Kälin: "If this all works out, it will be possible to import clean wool from New Zealand's Climatex Club and use it in our biologically cycleable Climatex products. After usage we collect the waste, use the energy, and bring the fertilizer back to New Zealand. We will change the whole business model and prove that challenging existing systems can be successful."

Finally Kälin started to develop a new idea based on the idea of "moral imagination." He did this with the University of Virginia. He realized that when dealing with environmental innovation, one cannot work according to the same parameters as one was trained in before. People have to be encouraged to envision what could happen as a result of the decisions they are taking. It was a new way for Kälin to open more space for his employees and push them even further on the route toward sustainability by looking to the consequences that might result from actions.

This phase was characterized by the establishment of mature operational concepts and practices at Rohner that were then transmitted to others. That transmission was not just market development in terms of presenting customers with Rohner and its

Climatex range, but it also involved new markets where know-how was transferred in the form of licenses. Finally, it involved the promotion of ideas on sustainable design principles that were passed to others actor through a variety of networks.

3.5 *2005–2011: Confronting a Changing Market*

Five years after the launch of its Climatex range of products and after the market recovered from the lack of demand of early years of 2000s, Rohner began to face increasing competition from manufacturers who sought to create products similar to Climatex. Although the competitor products did not reach the same product quality or environmental specification as Climatex products, they did offer customers a functional specification at a lower price point. Rohner continued to produce its range at the Belgach site until 2008 when it was bought by Gessner AG. Gessner offered the Rohner Climatex range a position in a stronger and more diversified range of high specification upholstery textiles and a more secure financial base. The production facilities at Belgach closed, and the machinery and production of Climatex was relocated. At this time, Kälin left the company to join EPEΛ.

Since 2008, the increased competition for the market sectors occupied by Climatex have caused Gessner to search for new product formulations that provide a wider range of functional qualities in line with what customers now want from fabrics of this kind. They are also looking to develop products at other price points. That is the main focus of the current work of the textile designer Elisabeth Nuenemann, who began work for Rohner before it was acquired.

The Rohner case reminds us that sustainable design requires continuous change and that competitiveness is an ever present driver. Rohner was a market leader and innovator, but its position was progressively eroded by competitors. Sustainability is not a fixed state, and environmental sustainability does not by itself guarantee commercial success. Rather, it implies continuous innovation and change that seeks to meet environmental, competitive and social conditions and demands. In addition, while sustainable design involves collaboration, it also faces severe competition. A key issue is knowing when and how to collaborate and when and how to compete, as well as knowing when and how to innovate and change.

4 Discussion

The previous section portrays an exceptional case. Between 1981 and 2004, Rohner undertook wholesale change in the way it thought about its products and production processes as well as its business model, including the way it was organized and managed. The case depicts a dynamic and complex story. It begins with a sense of mismatch between the company and its business, social and environmental context. This developed into the search for a new vision of what the company would need to become. This vision provided Rohner with what could be understood as a distant horizon for the company to aim for. That was supported by the development and testing of some novel concepts that linked the business of Rohner to the environmental and business concerns the business and its products faced. These novel concepts provided a more elaborated near horizon that could be then linked to ideas for change contributed by actors inside the company and within the supply chain of which it was part. These ideas provided the steps by which change was accomplished to make the concepts concrete and its new vision a reality.

Rohner's case unfolds a critical loop in the process of change. The ideas generated for change were dependent on and framed by the novel concepts and vision, which, in turn, were critical in shaping the direction of ideas for innovation and change. One of the factors that made this loop possible was the open communication approach and joint problem solving between the company and its managers, the CEO, and consultants. The open and collaborative approach was applied to concept development and the generation of ideas. One of the characteristics of this case is that learning and the search for solutions mattered more than hierarchies; ideas mattered as much as accomplishments; and the vision and concepts that provided stretch and ambition were more important than the comfort of stability.

The case also reveals the importance of shared leadership. Albin Kälin clearly championed the whole process but did not act on his own. The CEO acted with others to create the setting within which that sharing happened. Kälin placed value on the people and their contribution and sought to make them feel comfortable with change. He genuinely saw problems as challenges and opportunities, and saw personal leadership in terms of pushing the boundaries of convention. He recognized that to push those boundaries he would need inputs from many others. He set out to align what he said with what he did, and with how he did it and he expected that of others. He assumed the position of a role model, setting the climate for others to provide the inputs required for Rohner to move from personal and shared leadership to organizational leadership and innovation in terms of technologies, products, know-how, and know-how to innovation.

A third key finding of this study is about the role of networks. The case suggests that Rohner progressed through processes of learning, innovation, and change in three separate yet connected networks. The same core ideas about the need for new vision, the development and deployment of novel concepts that linked environmental and business performance and the contribution of ideas was common to the way Rohner interacted with these three networks. These are shown below in Fig. 7.1.

The first network was the upstream network of suppliers of materials that contributed to the products that Rohner made, the designers of a new layout for the mill and for new elements in the physical infrastructure of the company, such as the dyehouse, and its product structure. This industrial network held the key elements that would be needed for Rohner to make physical progress in the redesign of its portfolio of products for sustainability. This network was somewhat resistant to change, but the possibilities could be found within it to enable change to take place.

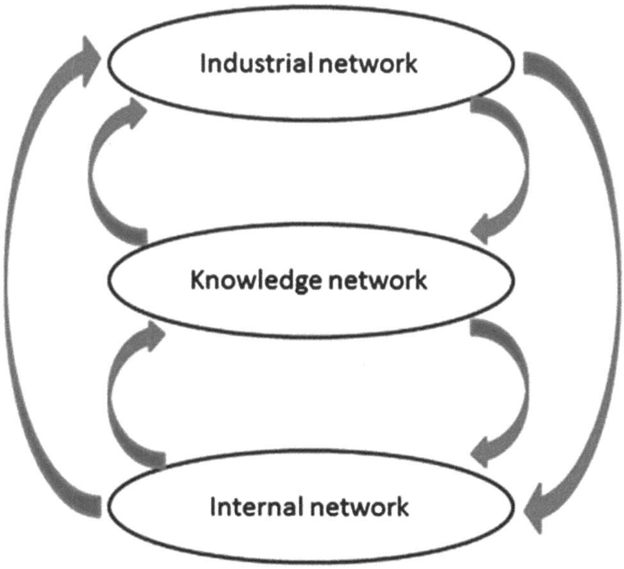

Fig. 7.1 Networks interrelations

By 1996, new product developments fed into the downstream network of customers that had begun to search for quality products with added-value around sustainability. Having worked with suppliers at the front-end of the move toward sustainably designed products, Rohner was positioned to offer these to the customers. Rohner did not first look for a market opportunity but for products that combined functionality and added value.

These achievements in the industrial network required Rohner and Kälin to operate in a knowledge-network that brought in new ideas on which the core concepts that informed Rohner's innovation processes and redesign were founded. Radical environmental thinkers, who were pushing the frontiers of sustainable development as a basis for design, were used to fashion concepts that would work at Rohner. These people, including William McDonough and Michael Braungart, were instrumental and simply part of a network of environmental thinkers and designers who had been exploring radical concepts needed for environmental management and sustainable development. They were linked through ideas to the environmental movement in the USA and Germany and to the global sustainable development movement. McDonough and Braungart played an important intermediary role, carrying these ideas for the network of environmental thinkers to a new audience at Rohner. What was rather less known at that time was how to make these ideas a business and organizational reality. Today, these ideas now reach a wider audience through their book (McDonough and Braungart 2002).

Rohner's case shows how this process of knowledge exchange through the network happened. But interestingly, the case goes further. Rohner's experience in operationalizing the concept of product stewardship and close-loop production consumption systems as a basis for a business model was brought to the agricultural community in New Zealand with the objective to find champions of those new concepts within new communities. In that sense, they were extending the knowledge network by bringing ideas practiced at Rohner to new audiences (farmers) and through that creating new communities of practice (the Climatex Club) with its commitment to examine what sustainable design and innovation might mean for them and their industrial network. In this way Rohner acted as a follower and learner in the original knowledge network of sustainable design and then became a leader and tutor in the extension of that network to farmers in New Zealand. At each stage where the network found a group of actors that wanted to explore the ideas in a concrete way, a community of practice formed itself. McDonough and Braungart served as a community of practice; Rohner was another community of practice; and so too was the Climatex Club in New Zealand. The original network of knowledge found new applications through these communities of practice.

Finally, a third type of network can be identified. The knowledge network that focused around Rohner and created a community of practice founded by the combination of McDonough and Braungart knowledge and Rohner's experience in textiles began to develop new concepts. This community of practice was supported by the ideas-action network within Rohner. The community of practice formed in Rohner relied on inputs from the staff as they would need to change their routines and practices to ensure that products and processes were designed and made in a way consistent with the new concept. Rohner had more control over this network because it was made up of its workers and associates. Even though Rohner had some control over the individuals in the ideas and action network, the managements' approach owed more to collaboration than control.

There were strong similarities in how Rohner worked with these three networks. Progress followed a particular sequence—vision before concept, concept before ideas and actions, with testing and exploration at each step, changing product structure came before market. These processes were supported by beliefs that informed the way people interacted. And there were certain roles played by different actors that contributed to learning, innovation and change. These have been observed in other companies that were also exceptional in their own way (D'Amato and Roome 2009). These process elements, combined with leading-edge ideas, provided for the change at Rohner and within its wider setting.

5 Conclusion

The exceptional case of Rohner supports the hypothesis with which this chapter began. Progress toward sustainable development implies organizational leadership to effect learning, innovation, and change in a focal company and its associated networks of actors. This means that progress toward sustainable development involves multilevel change based on technological and managerial innovation that effects both organizational and social change.

The chapter goes further by suggesting much about the nature of individual and organizational leadership, the processes by which learning, change and innovation take place and the critical role of networks of actors that engage new actors and form new communities of practice as part of the process change. It includes networks that span organizations linking them by material artifacts by knowledge and by a shared concern to deploy ideas in a common venture. The chapter shows that the beliefs and the roles that are assumed by individuals and groups in these processes are key to create a climate for change and design for sustainability. It reveals that this is a sequenced process.

The Rohner case study also puts forward that the process for innovation and design for sustainable development challenges the concept of "open innovation" that has recently found much currency in companies interested in innovation. An open innovation approach refers to an innovation process in which the boundaries of the firm are porous using inflows and outflows of knowledge to accelerate internal innovation and expand the markets for external use of innovation (Chesbrough et al. 2006). This concept of innovation encourages firms to use external and internal ideas. Indeed the case suggests that innovation for sustainable development requires a process that is even more open than that envisioned in the literature on open innovation and that sustainable development is best understood as a management innovation in innovation. The reason for this extreme openness in innovation for sustainable development is that it involves a focal business organization, other business and nonbusiness organizations, and a range of social actors in innovation and change. The points above have serious implications for practitioners in business and policy-making for sustainable development, for those engaged in innovation, strategy, organizational change, and leadership, and above all else for those involved in the education of future managers, particularly business schools.

The case leads to a number of questions that need to be further researched. Multicase studies could help to identify whether the process, practices, beliefs, and roles identified through Rohner are specific to this case or can be generalized to companies engaged in profound change. It would also be valuable to explore whether or not the same approaches found in innovation and design for sustainability can also be found in other types and fields of innovation. The questions to examine relate to the specificity and uniqueness of the innovation process for sustainable development. Some research already suggests that innovation driven by sustainable development is qualitatively different and requires different competences and capabilities from innovation that is directed to competitive performance alone (Roome and van Kleef 2007). Further research will need to be careful to distinguish what attributes make innovation really "innovation for sustainable development." The authors note that many companies make claims about their commitment to sustainability, yet little is provided to substantiate why a practice or innovation can carry the claim of sustainability. It is possible that the practices described in this case, around design for sustainability as a purposeful multiactor process involving organizational and social change brought about through networks of learning and innovation, provide the basis for a definition of sustainable development in the business context. If so, then much of what is currently labeled sustainability in business would not pass that test. Indeed, very little currently described by scholars as terms of business and sustainable development would meet that test.

The final comment brings forward the implication of these ideas for the future of management education. Management education is particularly important because it contributes to the development of skills and attitudes that are used by managers in business. In future those skills will need to play a role in promoting and supporting organizational performance and sustainable development. This raises a question about whether the approach to sustainable development observed at Rohner, as a process of design involving learning and innovation based on a set of practices and beliefs, is yet well understood and taught in business schools or in management development programs? The evidence from existing teaching cases on Rohner suggest that the perspective taken in this chapter is rather unique, even though it is consistent with sustainable development as understood for some 20 years. That perspective identifies the processes and the dynamic capabilities that played out around Rohner as it contributed to sustainable development through the action-learning networks of which it was part. If the Rohner case is about designing for sustainable development and the core elements of this case are not currently taught in business schools in courses on strategy, innovation, corporate responsibility or sustainable development, then we have a problem if we want to secure a more sustainable future. That would imply the knowledge network observed around Rohner has still to influence business school academics and future management practitioners and business leaders.

Acknowledgment This case study was part of a project originally conducted under the Dutch National Research Program on Corporate Social Responsibility (2003–2004) financed by the Ministry of Economic Affairs and coordinated by Dr. Jan Jonker and Professor Dr. Nigel Roome. The authors acknowledge the contribution of Marcel Postema and Robert-Paul Doove in the research and development of this case under the direction of Nigel Roome.

References

Agenda 21 (1992). United Nations, New York

Brundtland Commission (1987) Our common future. Oxford University Press, Oxford

Carson R (1962) Silent Spring Houghton Mifflin, Boston

Charter M, Tischner U (2001) Sustainable Solutions: Developing Products and Services for the Future. Greenleaf Publishing, London, UK

Chesbrough H, Vanhaverbeke W, West J (eds) (2006) Open Innovation: Researching a New Paradigm Oxford University Press

Clarke S, Roome N (1999) Sustainable business: learning – action networks as organizational assets. Business Strategy & the Environment 8 (5):296–310

D'Amato A, Roome N (2009) Toward an integrated model of leadership for corporate responsibility and sustainable development: A process model of CR beyond management innovation. Corporate Governance: The International Journal of Business and Society 9 (4):421–434

Eisenhardt KM (1989) Building Theories from Case Study Research. Academy of Management Review 14 (4):532–550
Gladwin T (1993) The Meaning of Green : A Plea for Organizational Theory. In: Fisher J, Schot J (eds) Environmental Strategies for Industry : International Perspectives on Research Needs and Policy Implications. Island Press., Washington, D.C.
Gorman ME, Werhane PH, Mehalik MM (1997) Rhoner Textil, AG (A, B, C, and D). University of Virginia
Gorman ME, Werhane PH, Mehalik MM (1998) Rohner Textil (E). University of Virginia
Hall J, Vredenburg H (2003) The challenges of innovating for sustainable development. MIT Sloan Management Review 45 (1):61–68
Interface (2011) Toward a more sustainable way of business. Accessed 26 May 2011
Kälin A (2001) The development of Climatex® Lifecycle™: a compstable, environmentally sound upholstery fabric. In: Charter M, Tischner U (eds) Sustainable Solutions: Developing Products and Services for the Future Greenleaf Publishing, London, UK, pp 393–401
Khan SA, Steger U (1999) Rohner Textil AG (A&B). IMD-2-0077 and IMD-2-0078,. International Institute for Management
Kok W (2004) Facing the challenge: The Lisbon strategy for growth and employment. European Community, Brussels
McDonough W, Braungart M (2002) Cradle to cradle: Remaking the way we make things. North Point Press, New York
Patton MQ (2002) Qualitative Research & Evaluation Methods
Pettigrew AM (1990) Longitudinal fild reserch on change: theory and practice. Organization Science 1 (3):267–292
Roome N, Bergin R Action-learning Networks and the Sustainable Enterprise: The Case of Ontario Hydro. In: Change SoMO (ed) American Academy of Management Meeting, San Diego: California, Aug. 9–12., 1998
Roome N, Bergin R (2006) Sustainable Development in an Industrial Enterprise: The Case of Ontario Hydro. Business Process Management Journal 12 (6):696–721
Roome N, Cahill E (2001) Sustainable Production: Challenges and Objectives for EU Research Policy. European Commission, DG XII
Roome N, Clarke S (1995) Management for Environmentally Sensitive Technology: Networks for Collaboration and Learning. Technology Analysis and Strategic Management 7 (2):191–215
Roome N, van Kleef H (2007) Developing Capabilities and Competence for Sustainable Business Management as Innovation: A Research Agenda. Journal of Cleaner Production 15 (1):38–51
Roome N, Winn S (1993) Responses to the Environment: Current Theory and the Implications to Practice and Research. R&D Management 23 (2):147–160
Siggelkow N (2007) Persuasion with case studies. Academy of Management Journal 50 (1):20–24
UNEP (1976) Annual report of the United Nations Environment Program. Nairobi: UN
Yin RK (2003) Case Study Research: Design and Methods. Third Edition. Sage Publications, London
York J (2006) Rohner Textiles: cradle-to-cradle innovation and sustainability. University of Virginia.

Chapter 8
The BBVA Innovation Network

Álvaro Morón Alonso and Javier Sebastián Cermeño

Abstract Setting an example for the global financial industry, BBVA established its Innovation Network in 2006 to be an open innovation ecosystem. This chapter describes the origins, network selection of five international poles, and various development stages of the Innovation Network, as well as the planned future configuration. In addition, a case study of a new ATM machine named ABIL is presented, in which the goal was to create sustained value by enhancing each and every dimension of customer experience.

BBVA is a global group that offers individual and corporate customers a complete range of financial and nonfinancial products and services. BBVA holds a leading position in the Spanish market, where it first began operations over 150 years ago. Today, the financial company has a strong international presence—it has a leading franchise in South America; it is the largest financial institution in Mexico; it is one of the 15 biggest commercial banks in the USA; and it is one of the few large international groups operating in China and Turkey. BBVA employs 104,000 people in over 30 countries around the world and has more than 47 million customers and 900,000 shareholders.

BBVA's corporate vision is expressed in a single idea: "We're working for a better future for people". To achieve this vision, innovation is one of the key elements that differentiates BBVA from its competitors and drives its organizational and business development. This focus separates it from conventional banks and positions it

Á.M. Alonso (✉)
BBVA, Pza. Santa Barbara, 2, Madrid 28004, Spain
e-mail: alvaro.moron@bbva.com

J.S. Cermeño
Capgemini, Pza. Santa Barbara 2, Madrid 28004, Spain
e-mail: javier.sebastian-cermeno@capgemini.com

S.P. MacGregor and T. Carleton (eds.), *Sustaining Innovation: Collaboration Models for a Complex World*, Innovation, Technology, and Knowledge Management,
DOI 10.1007/978-1-4614-2077-4_8,

firmly en route to becoming a new class of industrial organizations specializing in the distribution of financial and nonfinancial services.

BBVA has the ambition and motivation to set new paradigms. BBVA's values are defined by seven simple corporate principles, expressed as follows:

1. Focus on the customer as the center of our business.
2. Creation of shareholder value.
3. Teamwork as the key to generating value.
4. Management style that generates enthusiasm.
5. Ethical conduct and personal and professional integrity as a way of understanding and conducting business.
6. Innovation as the engine for progress.
7. Corporate social responsibility as an intrinsic part of development.

1 The Importance of Sustainable Value in Innovation

BBVA's innovation model creates sustained value by focusing on each and every dimension of customer experience. BBVA believes that innovation is the driving force behind economic growth and long-term improvement in standards of living. Innovation is needed more than ever to tackle the enormous challenges facing the human race—including inequality, poverty, education, health, climate change, and the environment. Our societies and economies require huge amounts of innovation if we are to make widespread improvements to the wellbeing of almost seven billion people (and growing) compatible with preserving our natural environment for future generations.

This ethical commitment extends to all of the societies in which we operate and to global society as a whole. We make this commitment because we firmly believe that economic development and social stability are key to BBVA's sustained and profitable growth.

2 The Network's Role in BBVA's Innovation Model

BBVA's Innovation Center, located in a unique building in central Madrid, Spain, makes a major contribution to the organization's innovation model. One of the Innovation Center's key functions is to make innovation tangible. In line with this, the Innovation Center has evolved into a Living Lab, where knowledge and experience are shared, accelerating the innovation process and transforming ideas into reality.

Our innovation model is open, participative, and is committed to, supports and facilitates disruptive innovation. Under this model, the Innovation Network performs a dual function. First, it assists the company's Global Observatory in carrying out continual exploration, acting as a source of knowledge about emerging socio-economic and technological trends, a role in which it not only draws on the findings of its own internal lines of research, but also serves as a nexus with secondary

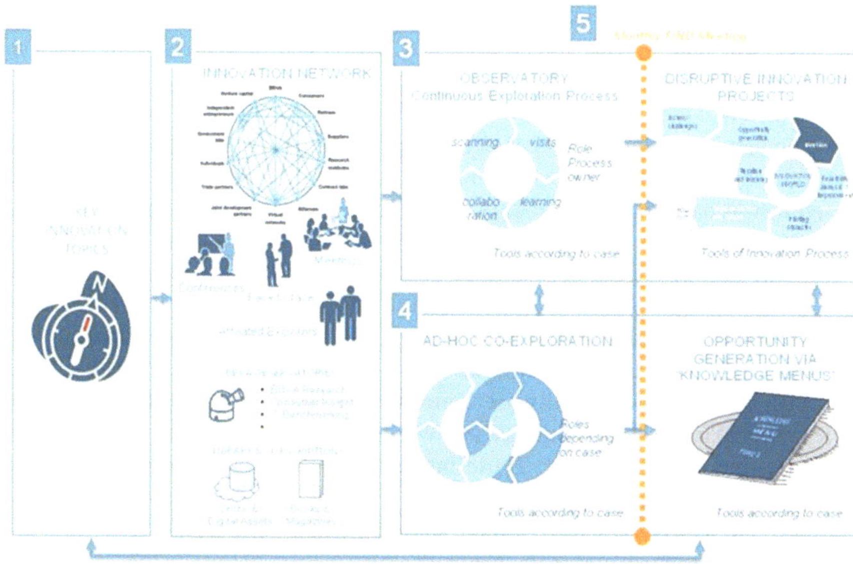

Fig. 8.1 The coexploration framework of BBVA's Innovation Network (© BBVA Group. Reprinted with permission)

sources in its network of contacts. The Global Observatory is a part of the Innovation Unit, responsible for the identification and following of socioeconomic, scientific and technological trends, and the finding of new business models and technologies likely to produce a disruptive innovation and potentially applicable to the bank. Second, the Innovation Network provides rapid responses to specific knowledge requests made by the bank, which are submitted via the Innovation Unit under the ad hoc coexploration framework, illustrated in Fig. 8.1.

Compared with the usual approach adopted by Spanish companies, in which innovation almost always comes from within the organization, opting for an open innovation framework constitutes a genuinely disruptive change. This approach is particularly novel in the financial sector, which, unlike other industries such as telecommunications or energy, is perceived as lacking creativity, despite being one of the Spain's most technologically advanced sectors.

Given this context, creating the Innovation Network was a challenging task, as no other company comparable in terms of either size or business model had attempted to do so before in Spain.

3 The Innovation Network's Origins

In November 2006, when the decision was made to create the BBVA Innovation Network, the idea of a financial institution developing an innovation network was received with a certain degree of incredulity not only within the bank itself, but also within the financial sector and in innovation circles in general. Consequently, the initial response from the research institutes contacted was lukewarm at best. It soon became clear that creating the network that BBVA had designed would be a long and laborious process.

The detailed design of the Innovation Network's structure had been drawn up in the months prior to this initial contact. Right from the start, the decision was made to work only with the best, with innovation's layer 1. Furthermore, as the model designed was based on the premise that the ideas received should derive from emerging initiatives (in the sense that they should not already be present in the marketplace), the network nodes needed to be institutions and companies that, on the one hand, were among the world's best known for their innovation capacity and, also on the other hand, had laboratories capable of putting those ideas into practice.

After analyzing innovation clusters across the globe, the decision was made to structure the Network around what were identified as the world's five innovation poles, located in the USA (both East and West Coast), Japan, Central Europe (essentially Germany), and Finland. These innovation poles were selected as they are responsible for a huge percentage of the patents registered worldwide every year and have the added advantage of representing a variety of innovation cultures and processes, all of which enrich the Network.

The following criteria were applied when selecting the innovation poles:

1. Structural innovation conditions.
2. Volume of investment in R&D.
3. Corporate-level innovation effort.
4. Number of patents registered.
5. Growth of Summary Innovation Index (SII), which gives an at-a-glance overview of aggregate national innovation performance[1].

After assessing which research institutes were traditionally considered to be leaders in the field, it became clear that the following should be prioritized:

- Massachusetts Institute of Technology (MIT), as the nexus of the innovation pole on the US East Coast.
- Stanford Research Institute (SRI), as its equivalent on the US West Coast.
- Instituto Fraunhofer-Gesellschaft, Germany, as the nexus of the innovation pole in Central Europe.

Therefore, these were the first institutes contacted. In each of the aforementioned innovation poles, the relationship was bolstered by contacts made with a series of companies identified as innovation leaders, among them IDEO, Intel, Microsoft, IBM, HP, Continuum, NCR, and Philips.

Appropriate contacts were then made in Finland, starting with the Otaniemi technology cluster, VTT (the Technical Research Center of Finland), and Tekes (the Finnish Funding Agency for Technology and Innovation), and companies such as Nokia. Finally, BBVA sent an initial learning expedition to Japan to make contact

[1] See more details at http://www.proinno-europe.eu/page/summary-innovation-index-0.

with research institutes, such as the AIST at the University of Tokyo, as well as with numerous firms, among them Fujitsu and NTT DoCoMo.

The network effect produced by this first round of conversations enabled the Network to expand relatively quickly and, consequently, to seek rapid results that would validate the applied knowledge development process designed.

4 Current Configuration: Five Innovation Poles

Right from the start, the Innovation Network has enabled BBVA to share information with specialized science and technology sources, with experts from various institutions and fields, and with companies from around the world. As mentioned above, to ensure a clear focus, BBVA identified the five global innovation poles where innovation processes were most highly developed and concentrated. All of the key players in these poles, ranging from universities to entrepreneurs and companies, judged to have the capacity to add value to the Network were identified.

This selection has resulted in an open innovation ecosystem comprising universities, institutes, governmental organizations, leading technology firms, investors, entrepreneurs, and technology incubators. Today, the Innovation Network is a fully developed and stable open innovation ecosystem spread across eight countries on three continents. It consists of over 40 institutes and technology companies, as well as numerous universities, governmental organizations, investors, entrepreneurs, technology incubators, applied research institutes, companies working with emerging technologies, and several highly renowned science and technology sector analysts.

Having successfully completed the first two development phases (which involved creating the Network's core and then consolidating and stabilizing it), the Network has now started to expand into emerging countries such as Brazil, home to some of its newest members.

In addition to maintaining constant contact with all of the Network nodes, several members of the Innovation Unit's Observatory regularly visit partner institutions and companies to strengthen professional ties with them, assess their laboratories, receive first-hand information about the latest advances in their lines of research, and develop joint projects.

The relationship between BBVA and the MIT is highly illustrative of the type of partnership the bank establishes with Innovation Network members. This relationship started in 2007 when, following the initial contact made, BBVA became the first European bank to join MIT's Industrial Liaison Program (ILP). Since signing the ILP agreement, BBVA has participated in several of the MIT's technology development and management programs.

Moreover, in 2010, the bank took its relationship with the MIT a step further by joining the MIT Media Lab's Digital Life Consortium and signing a Consortium Research Sponsorship agreement. Under this agreement, a BBVA representative has joined the Media Lab team. BBVA has also since signed agreements with MIT Sloan's Center for Information Systems Research (CISR).

What assessment can be made of the Innovation Network's first few years in operation? Since being set up, the Network has enabled the bank to identify the socioeconomic, scientific, and technological trends that will shape the future and has allowed it to work with some of the most innovative companies and people on the planet, developing disruptive projects that are now being implemented in BBVA. ABIL, the bank's new ATM (automated teller machine), is a good example of the latter.

5 A Case Study of ABIL

In 2010, BBVA finalized testing and began piloting its new ABIL ATM in selected bank branches. Development was made possible by cooperation with multidisciplinary companies such as IDEO, Fujitsu, and NCR, all of which are members of the Innovation Network.

The traditional product development and relationship models in the ATM industry have changed, with user experience taking precedence over the financial sector's need for automation. The ABIL product was developed openly, and access has been granted to the rest of the industry rather than keeping it exclusively as a proprietary BBVA development. ABIL is the result of an exhaustive study into customer behavior and is designed to offer the full service that meets the needs of both regular ATM users and nonusers. ABIL is based on the three key premises that the system should be simple, human, and flexible.

Simplicity is an essential feature and is achieved by employing a vertical touchscreen with large buttons. ABIL has just one multipurpose slot, via which it receives and dispenses the banknotes, receipts, and other documentation required in the various operations performed. This is known as the "One Slot concept".

ABIL is flexible. It remembers customers' most commonly performed operations and streamlines usage by providing direct access to them. Users can also work with various associated accounts via the ABIL ATM. In addition, when withdrawing money, customers can select the banknotes' denomination.

One of the biggest innovations is found in the external design of the ATM, which is shielded by a vertically positioned screen. This allows users to access their accounts without being overseen and makes ABIL both more human and more private than conventional ATMs. This new design also facilitates access for disabled users. In addition, the ATM includes a wide shelf where users can leave their handbag or briefcase whilst using the ATM. ABIL also provides helpful nonfinancial information, such as weather forecasts.

This design, shown in Fig. 8.2, won BBVA The Banker's Innovation in Banking Technology Award 2010 in the Delivery Channel Technology category. The Banker Awards annually pay tribute to banking technology excellence and innovation. Since it was founded in 1926, The Banker, which is owned by the Financial Times Group, has been a highly respected source of financial information. Its annual accolades highlight the ways in which financial institutions use innovation to create solutions, services and strategies to deal with business challenges.

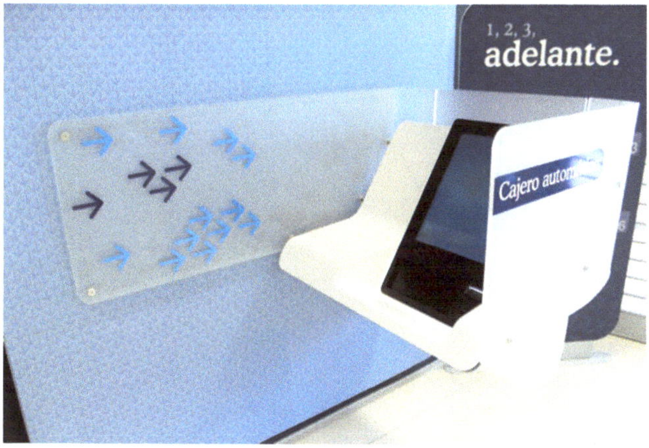

Fig. 8.2 Design of BBVA ABIL ATM (© BBVA Group. Reprinted with permission)

Recently, the MOMA selected ABIL's interface for inclusion in its "Talk to Me" exhibition held this summer.[2] ABIL has also been selected for the iF Communication Design Award, a highly renowned design trophy.[3]

6 Looking to the Future: Innovation Network Expansion

The Innovation Network is now performing well, presented in Fig. 8.3, and has achieved varying degrees of consolidation in its various poles. And, as has been described, the ecosystem is starting to produce tangible results.

Furthermore, the Network has helped BBVA identify other nontechnological forces that are going to transform the financial industry and society in general. In order to identify the opportunities that these new scenarios offer, the Innovation Unit plans to expand the Network along the following lines:

Innovation: In addition to the five primary innovation poles identified during the Network's initial design, several others were detected and earmarked for exploration at a later date. These poles have continued to evolve and other new ones have since emerged. Therefore, one of the Network's natural means of expansion is to develop relationships in these emerging locations (currently led by South Korea and Israel).

---

[2] See more details at http://wp.moma.org/talk_to_me/whos-talking/objects/.

[3] See more details at http://www.ifdesign.de/beitragsdetails_e.html?offset=0&sprache=1&award_id=227& beitrag_id= 69571.

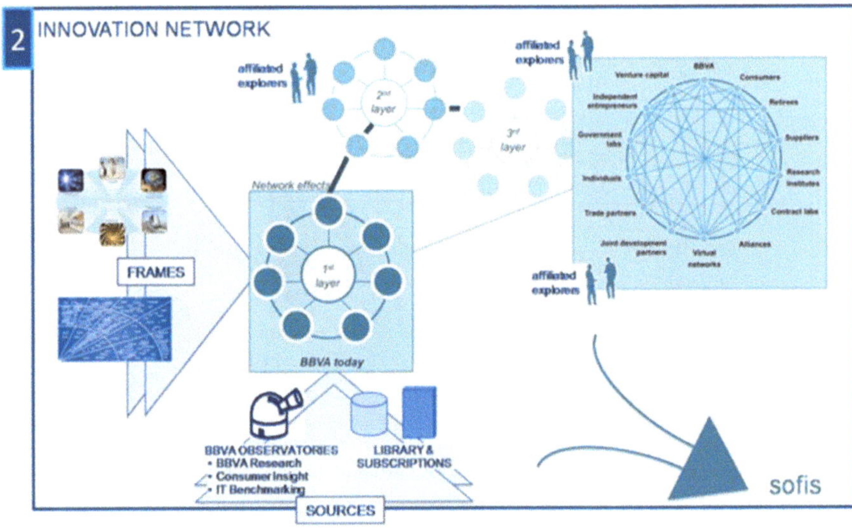

Fig. 8.3 BBVA Innovation Network expansion (© BBVA Group. Reprinted with permission)

Economic growth/geography: Several emerging countries are experiencing economic growth well above that of developed countries and are positioning themselves to become major powers in the near future (such as Brazil, India, China, and Turkey). In them, research and entrepreneurial communities, bolstered by strong economic prospects and actively favorable government policy, are producing significant amounts of innovation. BBVA's Innovation Network has to take these into account, which is why another of the development lines will focus on innovation leaders in each of these countries. In fact, the Network's first nodes in Brazil have already been identified and learning expeditions have already traveled to the other countries mentioned.

In later stages, the aim is to expand the Network to include those countries that currently represent the next wave of economic growth, several of which, among them Australia and Indonesia, already merit attention.

New sectors: Ever since the Network was first created, one of BBVA's aspirations has been to include, in later expansion phases, companies that do not focus solely on the technology or financial sectors. The bank firmly believes that the benefits of innovative experiences in other sectors apparently unrelated to banking can be transferred to BBVA, where they could then develop into disruptive innovation projects. In line with this, BBVA plans to expand the Network to include innovation leaders in sectors such as retail, telecommunications, manufacturing, energy, etc.

Financial services best practice: One of the most direct sources of ideas for innovative projects is the best practice by other members of the financial sector. In many cases, these practices are found in players in countries not included in the Innovation Network, but which nonetheless offer the potential to develop partnership frameworks that lead to valuable innovation projects for both parties. The focus here is to

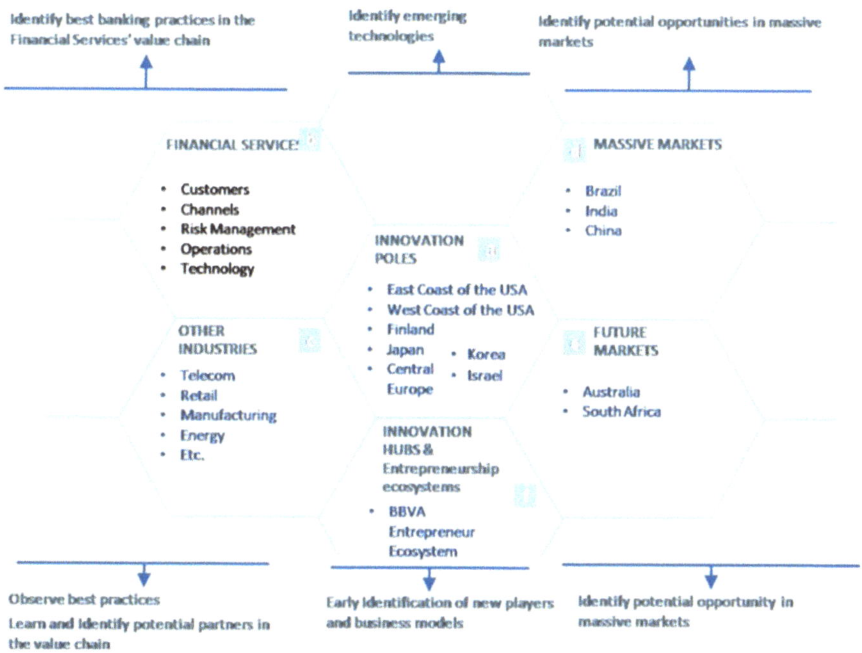

Fig. 8.4 Future framework structure (© BBVA Group. Reprinted with permission)

identify especially creative players with a view to developing the financial services of the future.

As the Network has developed, it has become clear that the new challenges are to (a) extend the network's reach, capacity, and coverage, (b) communicate innovation's values and message throughout the BBVA Group, and (c) encourage cocreation both inside and outside the Group.

To meet these challenges, BBVA has created a new role, that of affiliated explorer or scouter. Any Group employee can extend and strengthen the Innovation Network. Successful node expansion will depend on the scouter's motivation and commitment to exploration.

When the Innovation Network enters its next development phase, it will become a central node and will act as a nexus to facilitate knowledge exchange between network nodes, shown in Fig. 8.4. This function will be performed by holding events in both the physical and virtual realm attended by representatives from the entire network, by members of BBVA's Innovation Unit, and by participants in the internal entrepreneurship network also defined within the model. Under this approach, all of the individual components will form part of a single global and fully integrated innovation ecosystem.

The reason for extending the Innovation Network is to broaden the knowledge capture and observation tools' reach to identify new opportunities that will take BBVA beyond the limits of financial sector conventionality and make it one of the forces driving the transformation of the sector.

In conclusion, we have learned about sustaining innovation, since the decision was first made to construct the network in late 2006. It has been a long and exciting journey, not only across different geographies but through diverse innovation cultures too. We have found new ways of thinking, discovered surprising ideas, and met highly creative people. And this is only the beginning. We envision a future in which the innovation network is a global, interconnected, adaptative, and learning structure, a true planetary brain.

Chapter 9
A Network of Networks: The Case of the UK Technology Strategy Board

David Coates

Abstract In this chapter we describe how through the work of Technology Strategy Board, the UK's innovation agency, the UK Government's intervention has delivered a manifestation of the innovation quadruple helix in facilitating a Web platform that provides linkages between academia, business, and government departments. This is the story of a journey where the goal was to establish an interactive, open innovation network of networks, and where the start-point was a collection of siloed communities funded by the UK Government to represent a wide range of technologies and sectors. Originally, these communities were set up as pillars of excellence with a view to creating collaboration spaces for academics to share knowledge within their disciplines, but as time has progressed, stimulated by the need to identify routes to exploit this knowledge, businesses and government departments were encouraged to join. We describe why we embarked on this journey, what was found at the start-point in these silos of excellence, and the challenges we faced in designing and developing the network of networks. Finally, we describe how, over the last 2 years, we have implemented an interactive, open innovation, challenge-led Web platform, inclusive of all who want to play.

1 Background

In these days of financial constraint and reduced government spending, we have an ever-increasing obligation to justify and demonstrate the impact from spending the public purse. This applies especially to areas such as support for soft activities, such as networking. In good times, monies are made available for networking in both

D. Coates (✉)
Growth Agenda Limited, 8 Church Street, Stapleford, London, UK
e-mail: david.coates@growthagenda.com

S.P. MacGregor and T. Carleton (eds.), *Sustaining Innovation: Collaboration Models for a Complex World*, Innovation, Technology, and Knowledge Management,
DOI 10.1007/978-1-4614-2077-4_9, © Springer Science+Business Media, LLC 2012

academic and business arenas because it is intuitively believed to be a good thing. However, today we have to prove this well beyond belief. The lessons we have learnt over the last 2 years are as a result of seeking out exemplars of the impact of networking, such as the number and value of start-up companies that have resulted from the serendipitous meeting of unlikely partners, to the need for networks to be technologically agnostic and neutral, to the extent that trust in, and value of, orchestrated interactions is never questioned.

We have extensively questioned business leaders about the value of the knowledge sharing networks in the UK, and we have been told resoundingly that neutrality is the main reason for continued engagement and interaction. This paradigm of neutrality has been facilitated by government support, both financially and through structural independence.

That notwithstanding, when a network becomes established and effective, the issue invariably debated is the potential to encourage self-sustainability, thus reducing financially dependency on government grants. This has been considered with models such as membership subscription being tabled, but whenever tried, erosion of neutrality has been observed followed by an outcry from subscription-based organizations, such as trade bodies and the professional institutions; the accusation being that the public purse is being used to distort the competitive landscape. Another concern has been that if payment if required to join, the principle of total inclusiveness is immediately negated.

In the UK, the holders of the public purse have accepted this argument, and grant funding continues for the immediate future.

Like many countries, the business base in the UK is made up predominantly of small to medium sized enterprises (SMEs). In 2009, there were over 4.9 million enterprises in the UK (Department for Business, Innovation and Skills 2011), of which 99.8% were less than 249 employees and 99.2% were less than 49 employees in size, in which only 6% (National Endowment for Science Technology and the Arts, NESTA 2009) are estimated to have the potential for high growth.[1] On the basis that large companies will have a limited capacity for innovation due to their innovation budgets being under considerable pressure, the result is that when considering innovation as a pathway to national economic growth, the challenge is to mobilize this highly fragmented group of high growth-potential small businesses.

When attempting to mobilize these high growth-potential companies, a number of barriers need to be overcome. The most challenging is that, invariably, these companies do not have the time or resources to look further than their immediate market and have enough to contend with in developing their immediate business. Coupled with the assertion that collaborative working can expedite innovation, it is for this reason that networks with a remit to stimulate open innovation, both within and across sectors and disciplines, have been established.

---

[1] The OECD defines a firm to be high-growth firm if it has an average employment growth rate exceeding 20%/over a 3-year period and had ten or more employees at the start of that period.

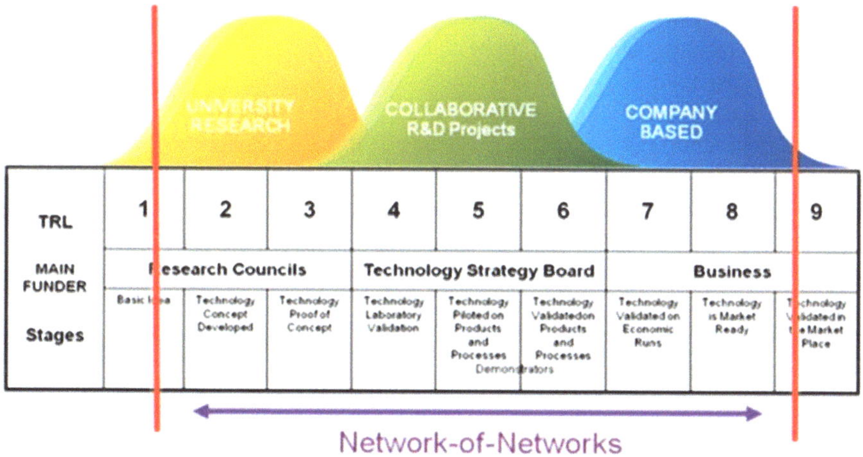

Fig. 9.1 Technology Readiness Levels.
Note: In the UK, there are seven Research Councils that invest around £3 billion in research, covering the full spectrum of academic disciplines from the medical and biological sciences to astronomy, physics, chemistry, and engineering, social sciences, economics, environmental sciences, and the arts and humanities. These councils are coordinated through the Research Councils UK

It has long been recognized that for innovation to be a key pillar for economic growth, it is essential that the knowledge base become fully engaged in the UK national innovation system. Taking the established US Department of Defense model of Technology Readiness Levels (TRL) (Wikipedia 2011), illustrated in Fig. 9.1 above, the national innovation networks have a remit that spans the whole TRL range, and to include academic input is an essential part of building the communities that constitute the network of networks.

The goal has been to develop a network of networks with a balance between the engagement with business and involvement of the knowledge base to facilitate the commercial exploitation of university research, whist maintaining the excellence of the basis for fundamental research.

In 2007, the UK government set up the Technology Strategy Board (http://www.innovateuk.org) with the objective of making the UK a global leader in innovation. To achieve this, the goals have been to ensure that the UK is a place where:

- Business is successfully competing at the forefront of technology and innovation developments globally.
- Government provides a supportive and coherent environment which allows innovation to thrive.
- Society understands, embraces, values, and is excited by innovation and technology.

In embracing the objective, the challenge has been twofold: to identify the businesses that have the potential to be classified as high growth,[2] and how to engage with these businesses.

There is no easy answer to this challenge, but the response has been to replace reactive intervention to the perceived needs of SMEs with creating an innovation climate to proactively encourage businesses to engage in addressing major societal challenges. For example, we have adopted a challenge-led approach to inviting small businesses to submit proposals for feasibility studies that address a major problem in the area of Assisted Living (Ansell 2010).

2 Innovation and Knowledge Sharing

2.1 *Why Knowledge Sharing?*

This may appear to be purely semantic, but we deliberately refer to *knowledge sharing* instead of *knowledge transfer or exchange*. In the case of knowledge transfer, the implication is that the transaction could be one way, to the benefit of only one party. In the case of knowledge exchange, knowledge could be traded for something other than knowledge, such as money as in the case of consultancy. Knowledge sharing is a process founded on mutual benefit.

2.2 *What Do We Mean by Knowledge?*

The premise is that data and information put in context becomes knowledge that can be shared. Knowledge is gained through context, experiences, and understanding. When the context is fully understood, it is possible to exploit the various relationships of experiences. As management expert Peter Drucker (1989) wrote once, "Knowledge is information that changes something or somebody—either by becoming grounds for actions, or by making an individual (or an institution) capable of different or more effective action." For example, a cookbook explains *how to* make a cake but an inexperienced cook, even with the recipe, might not make a good cake. However, it is only when the information in the cookbook is put in context of relevant experience and skill that, almost certainly, an excellent cake can be made.

---

[2] Note: in the NESTA report, the 6% represents high growth businesses employing ten or more people, accounting for some 11,530 firms in 2008. A comment about this report is that it does not take into account the start-up companies with less than ten people and the report does not mention how to identify this 6%.

3 Creating a Climate for a Knowledge Sharing Culture

Creating an effective knowledge sharing culture is about making knowledge sharing the norm. In creating this culture, there is a need to encourage people to work together effectively, and to collaborate and share—ultimately, to develop an environment where knowledge can and must be deployed and used more productively.

In developing this environment, a number of factors should be taken into account:

- Information must be put in context if knowledge is to be shared. This requires both the giver and receiver of knowledge to understand each other's situation and needs.
- The purpose of knowledge sharing is to help organizations meet their business needs and objectives; the overall goal must be to contribute to wealth creation and improving the quality of life.
- Learning how to make knowledge productive is as important as the sharing knowledge. This is one of the most difficult issues to be addressed. MIT scholar Michael Schrage (2000) wrote, "The ability to act on knowledge is power. Unfortunately, most people do not have the ability to act on the knowledge they possess."

The organizational tenet must be that "knowledge sharing is power," replacing the old paradigm of "knowledge is power".

Also, it is important to understand why the identification, acquisition, and productive use of new knowledge can be an essential part of business survival, particularly in a poor, unpredictable economic climate:

- The pace of technological, business, and social change is accelerating. As things change, so does the knowledge base erode—in some businesses, as much as 50% of what was the knowledge base of 5 years ago is obsolete today.
- Increasingly, the only sustainable competitive advantage is continuous innovation through the application of new knowledge.
- Intangible products—ideas, processes, and knowledge—are taking a growing share of global trade from the traditional, tangible goods of the manufacturing economy.
- With increasing turnover and redeployment of people, there is no such thing as a job for life anymore. When people leave an organization, their knowledge leaves with them.

4 Open Innovation, Knowledge Sharing, and Networking

In the past few years, innovation has evolved into the practice of open innovation. Originally put forward by Henry Chesbrough, the concept has now passed into every-day business life, with the boards of many companies appointing directors to act as promoters and guardians. Chesbrough et al. (2008) defines Open Innovation as "...the use of purposive inflows and outflows of knowledge to accelerate internal innovation, and expand the markets for external use of innovation... Open Innovation is a paradigm that assumes that firms can and should use external ideas as well as internal ideas, and internal and external paths to market, as they look to advance their technology."

Studies have shown that the companies successful with open innovation have provided simple interfaces and intermediaries, both inside and outside the organization, to manage the interactions, reduce the complexity of negotiation, create public space platforms for innovation, and invest in activities at the boundary of the organization (Institute for Manufacturing 2009). More fundamentally, as open innovation is essentially about linking external sources knowledge and routes to market with internal innovation activities, for companies to find success there has to be the means to communicate and work with people and organizations outside of their own sphere of experience; in many cases, accessing people and knowledge that they do not know they do not know.

With the rise in emphasis on open innovation, it is important to briefly touch on and understand the drivers for sharing knowledge. As early as the mid 1990s, it was recognized that there were many reasons for collaboration and knowledge sharing (Hagedoorn 1993):

In the areas of basic and applied research, and in some cases, technological development, the drivers for knowledge sharing have included:

- The increased complexity and cross-sector nature of the application of new technologies and the cross-fertilization of scientific disciplines.
- The minimizing and sharing of uncertainties and risks associated with R&D.
- The reduction and sharing of the costs of R&D.

In turn, these drivers can be related to innovation processes:

- The capture and sharing of tacit knowledge, facilitating technology transfer and the prospect of technological leapfrogging.
- The shortening of innovation cycle by reducing the timeline between invention and market introduction.

...and to market access, and the search for new opportunities:

- The gathering of insights about environmental changes and opportunities.
- To facilitate strategies for the entry to foreign markets.
- The development and expansion of products ranges.

Remarkably, very little has changed, and almost 20 years later, these drivers have become even more poignant. However, one of the main changes has been the realization that knowledge sharing, open innovation, and networking are symbiotic and go hand-in-hand.

Recognizing this symbiosis, over the last 2 years, the Technology Strategy Board has spent considerable time and effort in developing a network of networks inclusive of all the high technology sectors in the UK, identified as the portfolio of Knowledge Transfer Networks.

5 Evolving a Network of Networks

5.1 From Faraday Partnerships to Knowledge Transfer Networks

Typical of government initiatives, the concept of creating networks comprising government departments, businesses, and academics was very well intended but in no way coordinated. In the UK, central and local government departments and agencies, such as Business, Innovation, and Skills[3] (BIS) and the Regional Development Agencies, sponsored business communities to develop a proliferation of innovation and knowledge transfer networks (initially formed in 2004), and the Research Councils promoted the academic communities to share knowledge in-discipline through organizations, such as the Faraday Partnerships (Association of Independent Research and Technology Organizations, AIRTO 2001).[4] To the majority of people, the networking landscape was very confusing, extremely fragmented, and totally unsatisfactory.

In 2007, following the publishing of the *Sainsbury Review: Race to the Top* (Department for Business, Innovation and Skills 2008), which resulted in the formation of the Technology Strategy Board, the decision was made to bring some order to all these networks. This resulted in the birth of the Knowledge Transfer Networks (KTNs). Many were the former Faraday Partnerships with some set up to serve gaps in the business landscape, for example the Creative Industries KTN. However, in many cases, all that changed was the name, with little reference to the emphasis on the KTNs being designed to serve the needs of, and be directed by, business.

Initially, the fragmentation continued with the KTNs serving their respective communities in silos, with little cross-sector or discipline networking. In fact, with the exception of the KTN label, there was very little to show that the networks were in any way related. This is why, in early 2009, a Technology Strategy Board program was embarked on to optimize the performance of the KTNs with a view to establishing a network of networks, supported by a greater, cohesive presence on the internet.

The objectives of this program were twofold: to make these networks more accessible to high growth, technologically innovative SMEs and to promote open innovation through cross-sector, cross-discipline networking, in which business, academia, and government could work together to develop a world-class innovation ecosystem, both physical and virtual.

In optimizing the performance of the KTNs, one of the first activities was to merge some of the networks to enable those who had yet to join to understand the scope and relevance to their work. Initially, there were 25 KTNs ranging from Materials and Electronics to Digital Systems and Resource Efficiency. The merger resulted in 15 KTNs with a much greater remit to promote cross-sector networking and knowledge sharing.

---

[3] Previously BIS was known as the Department for Innovation, University and Skills (DIUS), and before that, as Business, Enterprise and Regulatory Reform (BERR).

[4] Faraday Partnerships were established between 1998 and 2003 to encourage closer contact and exchange between the science base and industry.

The 15 KTNs are:

- *Aerospace, aviation, and defense*—enabling innovation and collaboration across the aerospace, aviation and defense enterprise.
- *Biosciences*—serving the food, agriculture and biosciences sectors.
- *Chemistry Innovation*—driving innovation and value for the chemistry-using industries.
- *Creative industries*—the knowledge transfer network for innovators in the creative industries.
- *Electronics, sensors, and photonics*—to be a beacon for successful wealth creation in the plastic electronics, embedded systems, displays, lighting, instrumentation, control systems, and so on.
- *Energy generation and supply*—together accelerating a sustainable energy future.
- *Environmental sustainability*—promoting a sustainable future through innovation and knowledge transfer.
- *Financial services*—serving the banking and insurance sectors.
- *HealthTec and medicines*—building a healthy nation through business innovation.
- *Information and communications technologies*—creating competitive advantage in ICT for the UK by facilitating knowledge exchange.
- *Industrial mathematics*—exploiting the power of mathematics.
- *Materials*—bringing together the material supply chain to improve industrial innovation.
- *Modern built environment*—to increase the exploitation of innovation in the built environment for demonstrated business benefit.
- *Nanotechnology*—accelerating innovation in nanoscale technologies.
- *Transport*—seeking innovative business opportunities from greater intermodal transport integration.

Previously, for networks funded by a government department, guidance was remote, and the metrics were based on simple reporting statistics, such as numbers of members, numbers of events held, and the number and value of any interactions between members. This led to the behavior of driving for membership but unfortunately did not reflect the impact of how networking increased innovation capability or how networking contributed to revenue and growth. Over the past 2 years, we set out to change this behavior to one of encouraging people to join the networks and becoming active in stimulating and driving open innovation, dipping in and out of the KTN networking activities according to their personal and business needs. The model was based on the proliferation of social networks, such as LinkedIn, where it is noticeable that the number of people joining and the level of networking activity have both increased with the realization that LinkedIn can address, and satisfy, the personal challenges of remaining connected.

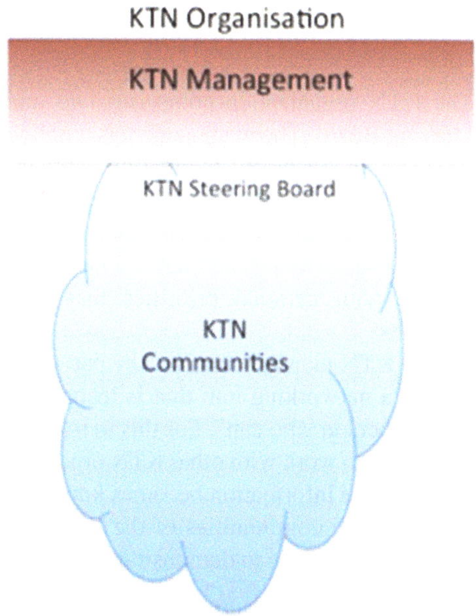

Fig. 9.2 Knowledge Transfer Networks (KTNs)

5.2 *What is a Knowledge Transfer Network?*

One of the legacies of the Faraday Partnerships is that a KTN has two components: the management that is responsible for the day-to-day running of the network and the communities that the network represents (Fig. 9.2). The relationship between the two is best described as a cloudy "T" shaped organization. This reflects the nature, and the challenge, of the two elements of the KTN—the communities which have a more narrow focus on knowledge held within the sector/technology area, while the management has a broader view of how knowledge should be communicated and shared between its own and other communities.

5.3 *KTN Communities*

Put very simply, the communities are companies of all sizes, academic institutions, research associations, government departments, and other organizations that make up or engage with the sector, or any source of capability relevant to these communities. As a whole, the collective of communities are both the source of and sink for knowledge when considering collaboration and knowledge sharing in the open innovation space. It is within these communities where the impact and outcomes of the networking and knowledge sharing activities carried out by the KTN are measured and reported.

5.4 KTN Organization

The KTN organization comprises two main elements: the Management and the Steering Board. The management is employed to run the network and is responsible for and orchestrates the facilitation of the innovative environment in which collaboration, sharing, and exchanging knowledge takes place. The KTN Steering Board is selected from business leaders in the represented communities, academics, and government departments, and has the task of identifying the themes and opportunities in line with national priorities that become the focus for KTN networking activities.

The KTN management is a key part of the portfolio of KTNs, and as such, operates as a networking hub that is technologically agnostic, where the mantra is "I know a person who can". For this to truly happen, the KTN management must have the ability to work with other KTN organizations in putting information into context such that the information becomes knowledge that can be shared to the mutual benefit of all the communities of the KTN portfolio. In working together, individual KTNs must be able to demonstrate certain characteristics:

- Acting as a translator—every sector, and even companies within a sector, has a specific business and technical language. The language is not only specific to the technology base and business but will also encompass the way companies communicate issues, problems, and challenges within. If knowledge sharing is to be truly effective across all sectors, then the KTNs must be responsible for translation and clarity of meaning when it comes to matching needs with solutions across the whole KTN network.
- Understanding and matching the "clockspeed" of differing sectors—every sector has a natural speed at which things happen. A word coined by Fine (1998), clockspeed refers to the rate at which companies and industries evolve. For example, fast clockspeed companies populate the information-technology sector. However, at the other end of the spectrum, heavy manufacturing, engineering and mining companies, as examples, exhibit much slower clockspeeds. Industry clockspeed is measured by the rates of change in product development, process creation, and organizational renewal. If cross-sector knowledge sharing is to be effective, then the KTN Management and Steering Boards must have an appreciation of respective sector clockspeeds and be able to interpret the significance of a mismatch of clockspeed on the uptake and use of new knowledge.
- Appreciation of the absorptive capacity of each sector, in which absorptive capacity is the concept used to measure a firm's ability to value, assimilate, and apply new knowledge (Bessant and Tidd 2011). It has been long since recognized that some industries have the ability to assimilate new knowledge and use this knowledge appropriately, whilst others have little appetite for anything new. The role of the KTN is to understand the absorptive capacity of the served

communities and to understand where the barriers lie and what needs to be done to overcome them. This is one of the first steps in creating a truly sharing community.

In developing the KTN network, the key to extensive knowledge sharing will be the ability of the KTNs to put information into the context of the issue or challenge being addressed as a way of establishing knowledge that can be shared effectively. The goal is to use the KTN network to increase profitability through developing innovative products and process in response to major societal challenges and hence raise the potential for the wealth creation of the UK.

In operating as part of the network of networks, a KTN does not replicate the work already being undertaken by specific trade bodies, industry associations and professional institutions; it provides a mechanism through which knowledge can be shared effectively in an inclusive way.

5.5 *The Network of Networks*

In developing the KTN portfolio, our goal has been to make it easier for inventive and innovative people to engage with the communities that can help realize their potential. By reducing the overlaps and the confusion in whom to talk to, we have developed a means of increasing the ability of all business, academic, and government communities in the UK to share knowledge in an innovation space where the network of networks has been designed to facilitate open innovation. In orchestrating this, it is the role of the KTN Management to establish the environment and culture of knowledge sharing and to facilitate the processes. In short, in facilitating knowledge sharing, it is the goal of the KTNs to moderate, translate, and communicate as an embodiment of the ultimate "semantic Web."

Figure 9.3 above represents the KTNs as the national network of networks. The challenge has been to engage with the UK SME business base, and in achieving this, to encourage the KTNs to work with regional and local organization such as universities and further education colleges (HEIs and FECs), university technology transfer offices (TTOs), research and technology organizations (RTOs), and iNets, which are regional networks funded to bring together local businesses in areas such as information computer technologies, health, and the creative industries. In addition to networking within the UK, and recognizing that innovation is not bound by national boundaries, we encourage the KTNs to network with international-facing government agencies, such as UK Trade & Investment.

To complement the physical activities of the network of networks, we have built a virtual open innovation environment called _connect (https://ktn.innovateuk.org/web/guest/).

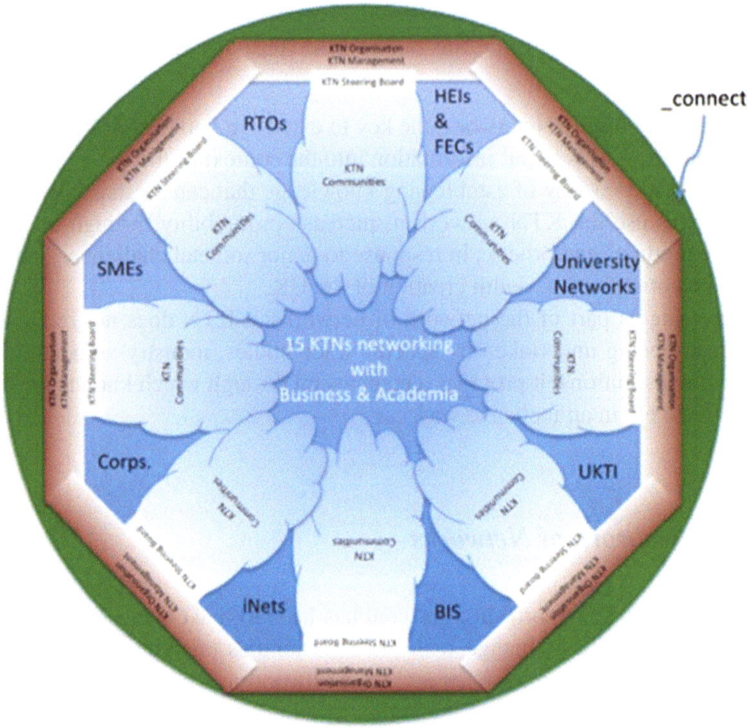

Fig. 9.3 Network of Networks

6 Connect: A Platform for Open Innovation

6.1 A Virtual Open Innovation Space

Fundamental to the success of this open innovation environment has been the development of the _connect Web platform, with the goal of creating a focal point in the UK for all networking and knowledge sharing activities, acting as an inclusive virtual space in which individuals from all communities can work together to explore open innovation. In parallel with this development, we have been encouraging the KTNs to evolve into behaving as a single network where communities from all sectors come together, where boundaries are broken down and the currency is the ability to freely collaborate to explore, enhance, and develop innovative new partnerships, products, and processes.

In the UK, _connect is becoming “the place to go” to be part of, and experience, open innovation; a virtual space where innovative people can meet to share knowledge and address challenges with the confidence that, as individuals, control of intellectual capital is not lost or intensions need to be publically declared.

_connect, as a free-to-use virtual space for open innovation, has the functionality and capability of matching challenges with the means of addressing these challenges. In doing this, the goal has been to attract innovative people with the notion that any challenge can be posted on the site which will then be matched with appropriate solutions; where challenges range from seeking collaboration partners and funding through to filling gaps in (technical) knowledge and finding new markets.

In addition to being the UK open innovation platform, since _connect has been developed with public funding, there is a need to demonstrate both financial and social impact. In this context, the expectation has been set that _connect will stimulate home grown entrepreneurship and encouraging investment in innovation.

To satisfy this expectation, _connect has evolved as a comprehensive communications platform where the Technology Strategy Board, as the UK's Innovation Agency, engages with individuals and businesses, where competitions are announced, submissions made, adjudication carried out, results posted, projects tracked and monitored, and impacts assessed.

_connect is the means for showcasing the range of interventions supported by the Technology Strategy Board in stimulating innovation amongst the businesses and academic communities who wish to engage in funded competitions: CR&D, KTP, SBRI, Grant for R&D, and EU programs. _connect represents an open communications system where discussions and debates take place, influencing future strategies, competition and other schemes with the objective of stimulating innovation. In short, _connect is at the heart of, and fundamental to, the Technology Strategy Board's cross business and academic engagement strategy.

_connect was launched on April 1, 2010, and after the usual teething problems, the 15 KTNs migrated their communities to the platform, and we continued to build functionality and improve performance.

During the early days, we recognized the need to concentrate on two main platform attributes:

- First, it is individuals who network and share knowledge, and that whilst we do need to track affiliation, the basis for the platform is that individuals must have an online home page that reflects their interests and needs, with alerts being sent when activities are initiated on the platform. To implement this, we introduced an extended sign-on, such that the more information volunteered by individuals about their interests and need, the more is fed to their home page.
- Second, for the platform to achieve success as an open innovation forum, housing the network of networks, as far as possible, community engagement should be all inclusive with any business, government, or academic community involved in innovation being encouraged to maintain a proactive presence.

Gradually the platform settled down and steadily grew to some 30,000 discrete individuals (20% academics, 18% international, and 75% representing SMEs) and 26 communities by April 2011. One major achievement was that whereas when the KTNs had their independent Web sites, only some 8% of the total combined community joined multiple networks, but with the single sign-on process for _connect, this figure leapt to 40%.

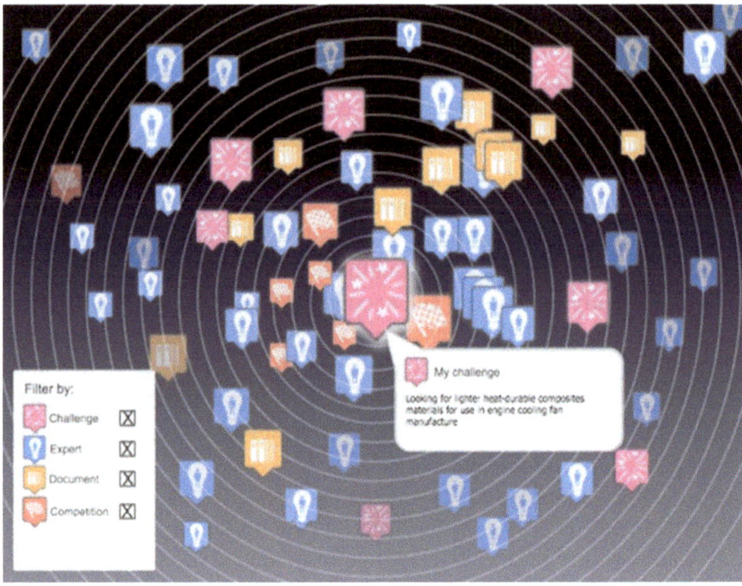

Fig. 9.4 Screen shot from _connectMe

Work progressed over the first year to build functionality and encourage other communities to become engaged. Our strategy was originally to establish the platform as the home of all 15 Knowledge Transfer Networks, and then bring onboard other communities, such as The National HE STEM and the Measurements Network. As the platform is developed, communities comprising the Research Councils and other government departments will continue to set up shop.

One of the final pieces of the puzzle is the functionality that makes _connect challenge-led. This has been delivered by the launch of the connect-me facility. This is based on the individual, within the spirit of knowledge sharing, being able to post "my challenge", be it seeking technology, people, funding or a collaboration partners, and appropriate resources are identified such that the "my challenge" is not seen by these potential resources but the individual can assess why the resource could be of assistance. This is shown Fig. 9.4 below, a typical screen shot where the distance from "my challenge" is an indication of the potential to help.

6.2 The Importance of Community Management

During the first year, we observed and experienced the extremes of success and failure of online community activity on _connect, from communities that thrive and grow to those that wither and die. In evaluating this widely varying performance, we conclude that, despite the functionality of the platform, the most important contributory factor is the quality of regular communication and engagement with, and stimulation of, the respective communities. In short, effective and proactive community management is the key to the successful growth and sustainability of any online community.

As a result of this experience, when introducing new communities, whilst we have adopted a policy of minimum prescription so as to act as a catalyst and not a key player in any interaction, we insist on complete oversight of all intended community management. Our goal has been to ensure that all the communities on the _connect platform have the appropriate management skills and resources to maintain a vibrant online presence.

In adopting this policy for community management, we have identified some basic rules for managing communities and, through identifying the attributes of the job of community manager, we have expressed the minimum we expect of communities when fully engaged with _connect.

Community management is the discipline of getting loosely federated networks of individuals to engage and work in productive ways. At a practical level, the role requires a unique set of skills that are very different from those commonly understood as management. For example, there is a requirement for an understanding of human behavior supporting the encouragement, promotion, and initiation of goal-oriented activities, without being prescriptive or controlling. Whereas, at a strategic level, the objective is to enable individuals to form symbiotic and energizing working relationships that generate mutual value, by bringing together disparate communities.

As part of the introduction of new communities, we assist with the planning of the community's approach to being part of _connect, help with hiring a community manager, provide the training for community managers, and create a methodology to measure community performance, and understand how the community fits into the larger business context.

7 The Future

Our goal is to build the communities that make up the network of networks to some 100,000 individuals by the end of 2011, to develop a truly open innovation platform. We will still need to ensure our primary focus is to support innovation in the UK, but since open innovation knows no boundaries, we will continue to develop international connectivity. One very clear objective will be to continue to encourage the use of _connect platform as one of the key vehicles for dissemination of academic research outcomes and to further develop as the focal point of the quadruple helix, drawing in players from the triple helix of government, academia, and business.

The proof of our success will be that _connect is the place to be, rather like LinkedIn and other social networking platforms, but with the difference that it is for the business community to share knowledge and access open innovation.

References

Ansell, C (2010 Dec 19) TSB HealthTech and Medicines: Assisted Living Innovation Platform (ALIP): Funding call to small and micro companies for feasibility studies in assisted living, UK _connect https://ktn.innovateuk.org/web/assisted-living-innovation-platform-alip/articles/-/blogs/tsb-funding-call-to-small-and-micro-companies-for-feasibility-studies-in-assisted-living
Association of Independent Research & Technology Organisations (AIRTO) (2001) The contribution of Faraday Partnerships to growth in innovation intensity in the UK economy http://www.airto.co.uk/pubpolpapers/2001_1policypaper.pdf
Bessant J, Tidd J (2011) Innovation and Entrepreneurship. John Wiley, West Sussex
Chesbrough H, Vanhaverbeke W, West J (2008) Open innovation: Researching a new paradigm. Oxford University Press, Oxford
Department for Business, Innovation and Skills (2011 Apr 15) Small and medium-sized enterprise statistics for the UK and regions http://stats.bis.gov.uk/ed/sme/
Department for Innovation, Universities and Skills (2008) Implementing "The Race to the Top": Lord Sainsbury's review of government's science and innovation policies. The Stationary Office, London http://195.88.100.72/resource/files/2008/12/16/racetothetop.pdf
Drucker P F (1989) *The New Realities*, Harper & Row
Fine C H (1998) Winning industry control in the age of temporary advantage. Perseus Books, New York
Hagedoorn J (1993) Understanding the rationale of strategic technology partnering: Inter-organizational modes of cooperation and sectoral differences. *Strategic Management Journal*, 14:371–385
National Endowment for Science Technology and the Arts (NESTA) (2009, October) The vital 6 percent http://www.nesta.org.uk/home1/assets/features/the_vital_6_per_cent
Schrage M D (2000) Serious Play: How the World's Best Companies Simulate to Innovate. Harvard Business School Press, Cambridge
Wikipedia (2011 Apr 7) Technology readiness level http://en.wikipedia.org/wiki/Technology_readiness_level
Institute for Manufacturing (2009) Creating an environment for open innovation. IfM briefing, 1(4). University of Cambridge, Cambridge http://www.ifm.eng.cam.ac.uk/working/briefings/v1n4_ifm_briefing.pdf

Chapter 10
Revolutionizing the Value Chain in the Sporting Goods Industry Through Virtualization

Dagmar Chlosta

Abstract The sporting goods market is a €59-billion business at wholesale. Add to that a fashion wholesale market of €125 billion, and you arrive at a sizeable industry that continues to build its value chain on traditional processes and technologies. This chapter discusses how a major sporting goods company, like many other consumer goods companies, faced the challenge of shortening product creation lead times to get closer to the consumer, reducing costs in the supply chain to improve the bottom line, and bringing its sales and marketing processes to the next level. An overhaul of the value chain was essential to meet the demands of the twenty-first century and delight consumers with a truly new brand experience. These challenges led to the creation of the virtualization strategy—a revolutionary, innovative approach to the value chain, unique in the sporting goods industry. The creation and implementation of virtualization became an innovation driver in cross-functional collaboration, speed to market, and supply chain efficiencies, and ultimately turned into a consumer stimulus.

The sporting goods industry is a €59-billion business at wholesale. Its products and product-related technologies are becoming more and more innovative and the consumers more demanding. The industry is on a continuous growth path. Sports and lifestyle are merging and becoming a new way of life. The fashion business alone accounts for another €125 billion of market share. However, the value chain and its related processes have remained largely unchanged over the years. The organization, processes, and systems follow tried and trusted patterns.

D. Chlosta (✉)
Adidas Group, Max-Reger-Strasse 187, Schwaig 90571, Germany
e-mail: dagmar.chlosta@googlemail.com

S.P. MacGregor and T. Carleton (eds.), *Sustaining Innovation: Collaboration Models for a Complex World*, Innovation, Technology, and Knowledge Management,
DOI 10.1007/978-1-4614-2077-4_10,

So imagine, your mission is to be the global leader in your field of play; you are determined to expand your business from traditional sporting goods into a more trend- and lifestyle-influenced area; you are on a continuous growth and expansion path, but your internal processes have remained largely unchanged over the years or even decades; and your value chain continues to be partially inefficient, slow, and costly. So what are your options, if you want to achieve your goals?

Of course, you could continue to do what you have done over the years and focus your innovation efforts exclusively on product technologies; after all, this is definitely one of your key strengths. You could also try to alleviate the pain points that your value chain imposes by introducing more resources to make up for inefficient processes and by tweaking the existing setup to make it a bit more palatable while helping you keep up with the immense growth that you are experiencing.

1 The Traditional Value Chain

What would this mean? What does this value chain look like? And what are the cornerstones of its collaboration model? If we take a look at the traditional value chain, it all starts out within the company's Marketing departments with the market and consumer research, which is then built into a seasonal concept that designers translate into product proposals. Then begins the process of product creation and prototyping. Another team in the organization, which resides in Operations, adds all the technical specifications to the design and sends the complete package off to the partner factories around the globe to start working on the first prototypes.

Fast forward a few weeks, and the first prototypes are sent back to headquarters (HQ), where the Global Product Teams consisting of Design, Marketing, and Operations staffs review the products and make needed changes; these changes are either based on commercial or design aspects of the prototype. This is where the process goes into round two—the changes are sent off to the partner factories, and another round of prototypes is created. A few more weeks pass by, and the review process starts anew. With a bit of luck, the second prototype round is the final one, but this is not necessarily a given one. Following internal alignment meetings, the product ranges subsequently are signed off, and sales samples for the markets are ordered.

In the meantime at the global factory base, everyone is busy booking and preparing materials for the sample production. Smaller batches have to be produced to ensure that every article in every color can be produced for the individual markets. The supplier network almost comes to a standstill in its regular bulk production due to the extremely high amount of sample production that is needed to service all the markets. This process involves a lot of manual work, a high coordination effort on the side of the Operations teams, and close control across the markets.

This is the so-called back-end process of the value chain, which is more or less a standard process across the industry. It is iterative, time consuming, and of course costly.

Looking at the front end, once the sales samples are ready for shipment, they are sent out to the respective markets and, depending on the different sell-in processes, are either displayed in central showrooms or shipped around the country and shown at specially prepared hotel rooms or major key account meetings at a central location. Regardless of what the individual market process looks like, it always requires setting up masses of physical samples for selection and ordering purposes. All depending on the range, some products cannot even been shown due to a sheer lack of space.

Moreover, physical samples are not just needed for internal review and external sell-in processes. Many samples are also needed for photo shoots for catalogue production, e-commerce, brand marketing campaigns, and other digital media.

All in all, both the back-end and front-end processes are extremely iterative, time consuming, and costly. They do not bring the organization closer to the consumer but rather remove it from being at the forefront of trend setting. Innovation, although at the core of product technologies, was not a key consideration for the internal process landscape.

From a collaboration perspective, this sequential process promoted silo thinking and a "throw-it-over-the-fence mentality," which entered another challenge into the equation. In the value chain, many internal and external parties play a key role. Their collaboration is key for the success of the company. It can almost resemble an orchestra: if one of the musicians plays out of tune, the whole composition is at stake. What happens if the players are not playing in sync? What happens if a silo mentality prevails? The value chain experienced exactly this challenge.

In order to create consumer products that are relevant and bring the consumer back for more, it is essential to closely connect the organization. However, in the traditional sequential value chain process, the different teams work in their expert silos and push information down the value chain to arrive at the subsequent result.

A batch process mentality existed at this major sporting goods company, and bigger picture thinking was hardly in place. If a designer created a product, he did not have all the information on hand to already design into profitability or engineering. This led to a high waste rate, as only in the later stages of the process products would be screened for their commercial application. External partners created materials and products based on information that was only as good as the designer and product manager that had created the brief. If the HQ team was not aware of the capabilities of the factories or lacked a view of the bigger picture, the creation of the prototype often resulted in less-than-optimal samples.

The challenge of the traditional value chain was varied with iterative processes, siloed teams, and high dependability on physical samples. All of these aspects were a hindrance to getting closer to the consumer and creating a value chain of excellence deemed fit for an industry leader. Therefore, staying with the old, traditional value chain did not seem a good option to ensure continued growth and entry into a more lifestyle-driven market.

How could the mold be broken? What was needed to truly think, act, and behave differently and to introduce innovation in all aspects of the value chain? Internal benchmarking would not be sufficient. Even more product innovation would not be sufficient. Tweaking the processes would not be sufficient.

2 The Concept of Virtualization

Why not take a radically different approach to the three cornerstones of the value chain: organization, processes, and systems? And why not look beyond our own industry for inspiration and innovation models? This is how the virtualization program was created—as a new way of thinking, a new process, a new technology; in short, a new value chain.

2.1 Introducing Virtualization

To introduce a game-changing innovation model, it was essential to look beyond internal processes and setups. It required building external partnerships with thought leaders that would not necessarily be linked with this industry.

With all the issues and challenges at hand, it was paramount to change the three cornerstones of business model: the organization, process, and systems. It was essential to become faster, more efficient, less costly, and more relevant while not losing sight of what had made the brand a leader in its industry, namely, product innovation, consumer relevance, creativity, and quality.

We set out as a small team of strategists and technical experts trying to find examples of other companies that were leaders in their field and how they had conquered the challenges of the value chain. We wanted to revolutionize the value chain by radically changing its cornerstones. No stone was left unturned to come up with a solution that would bring us all the needed benefits. This was the moment when the concept of virtualization was born. Our vision was to replace physical samples in the value chain with virtual ones. No more, no less. But how would we go about it?

2.2 Establishing Innovation Partnerships with External Parties

To achieve virtualization, a collaboration model was set up with a leader in the sector of 3D visualization, which had established a rapport in the automotive, aerospace, and consumer goods industry as the key innovation driver for 3D technology and processes.

As part of a strategic innovation project, we developed a technology and processes that would allow the virtualization of apparel, footwear, and accessories products in photorealistic quality. The joint journey in developing this technology for the apparel and footwear industry was made possible by moving outside of given parameters and boundaries and using external best practice models to learn, grow, and innovate.

While a robust network of internal experts existed, taking the given traditional processes and placing them in context with an external, very different world provided the actual competitive advantage. There was, of course, an initial surprise when we reached out to establish this partnership with the 3D visualization leader, but soon the new collaboration turned into an enriching and thought-provoking experience for both parties.

3 The Different Facets of the Value Chain

The value chain should not be considered a chain but rather a cycle because everything that we do starts with the consumer, and the consumer is ultimately also the recipient of the output. Therefore, all parts of the value chain from design to marketing, operations, IT, and sales need to share the same vision of how they want to win over the consumer.

3.1 The Product Creation Process

We first explored the product creation process in its different facets. After all, what had proven successful for one of the biggest projects in the aviation industry should not fail us in sporting goods. Why should it not be possible to link the traditional product creation of design, computer-aided design (CAD), and product lifecycle management (PLM) systems with external technologies and therefore make it more intuitive and open up a completely different universe of opportunities?

First and foremost, some technical difficulties had to be overcome on the footwear side. Apparel was pretty much covered through the existing technology by another third-party provider, and here it was essential to link the two external collaboration partners in a virtual network to further customize and improve the existing technology to make it truly industry leading. On the footwear side, the problem was somewhat more complex; there was no off-the-shelf solution in place. However, through intense collaboration between our internal experts and our external collaboration partner, we managed to create a technology that allowed us to create 3D virtual samples that were almost photorealistic.

This way, the back-end process of the value chain looked dramatically differently. No longer did we need a myriad of physical samples; no longer did we have to spend endless hours waiting for these samples to arrive for product reviews and finalization; no longer did we clog up the supply chain with our immense sample production. The process became a lot more integrated, efficient, and streamlined.

So now that we had the technology, now that we were able to provide virtual instead of physical samples, and now that we could actually revolutionize the value chain, the question arose how to drive this unique proposition into the organization.

3.2 The Sell-In Process

First, we had to clarify the scope and ultimate goal of our innovation project again. It was obvious—we wanted to save time to market in the product creation and sell-in process, save money in sell-in and brand marketing activities, bring the consumer closer to the brand, and use this unique innovative concept to lead a paradigm change in our industry.

If Rome was not built in a day, then virtualization was not implemented in a day either. We took it one step at a time. Coincidentally, at the same time as virtualization had become ready to be used, a major initiative was launched to reduce complexity in the organization. This proved to be the perfect vehicle to gain the needed momentum to implement virtualization in the sell-in process.

A project team was formed to tackle the complexity challenges of the sell-in process, and a key sales region was chosen to become the pilot for the replacement of physical with virtual samples in a key account and field sales process. In close collaboration with the internal IT department, we built a merchandising platform that would allow us to showcase the virtual sample next to the physical samples that would still be shown in the sell-in process around the globe.

A virtual network consisting of marketing, operations, sales, and IT team members was formed to first understand and analyze the existing sales processes, then establish a hybrid of physical and virtual sell-in processes, and subsequently create a rollout plan for the first launch season.

A key part of the rollout plan was the provision of the virtual samples. We had two choices: either create the samples in-house or collaborate with our sourcing partners and make them responsible for the provision of the virtual samples alongside the physical ones.

If we wanted to create the samples in-house, it would mean hiring resources in HQ in order to set up a virtual product creation team. Additionally, the virtual sample creation would be delinked from the physical one. This could pose many problems, for example a lack of comparability. We, therefore, opted for the outsourcing of this process to our sourcing partners. This meant that we had to put our sourcing partners in touch with the technology providers to establish a network, provide hard- and software, and conduct the needed trainings. In quite a few cases, it also meant hiring the right resources to have the necessary capabilities in place. This, in turn, rendered investments necessary that our sourcing partners had to make.

However, as the physical sample production had proven to be not just a problem for us but also for our factories, they were very open to move toward more innovative processes. After all, our sample production had presented them with many issues, such as extreme complexity and costs.

Besides the creation of the product files and the establishment of the new sell-in tool, it was paramount to understand how the sell-in process was conducted. We asked for detailed information of infrastructure, resources, setup, and customer behavior. The infrastructure part had us in for a surprise—little did we know that the showrooms that we wanted to equip with large LED screens, beamers, and Internet connection for file download were at times no more than freight containers that had been transformed into a "showroom." There were other examples, too: state-of-the-art showrooms with the latest and greatest in technology. However, we had to cater to every possible facet of the infrastructure.

3.3 Tradition Meets Innovation

Hence, our setup had to be easy, accessible, affordable, and intuitive. The plan was to provide one physical sample per style or model and display all other color ways in virtual form. This would allow the customer (i.e., retailer) to have the haptic product experience but would not make this experience necessary ten times over, if a T-shirt came in 11 color ways.

To better understand why this would potentially pose a challenge, it is important to know that almost all players in the sporting goods industry conduct the sell-in process with a full array of physical samples. This is what retailers are used to. They are used to merchandising the product in their showrooms on the floor or going through the complete collection with the sales person. Not having all samples available would be an unusual proposition indeed.

In order to ensure that the sales force would actually apply the new process and use the system, it was important to provide perfect color calibration and have an intuitive system in place that people actually enjoyed using. After all, the success of the project and the future of virtualization were completely dependent on how well-received the technology and process were with those using it and whether they would feel comfortable moving out of their comfort zone.

4 Internal Change Management: Making the Organization Part of Innovation

Major projects tend to fail not because of budget constraints or lack of feasibility. They do not fail because processes have not been developed with the needed due diligence or systems not in place. They fail because the "human factor" has not been considered; the organization has not been made part of the project, and the individuals are not clear what they are to gain from the change. They fail because the change is not properly managed.

4.1 Addressing the Doubts

While the project was progressing in the pilot region, while the virtual network was busy getting the technology ready, and while the core project team was eager to conduct the first launch, the HQ teams were less enthusiastic.

Doubts had come up from various parties that were not directly involved in the project that the project could deliver what had been promised. It was rumored that sales would be at stake, that retailers would never move away from demanding a physical sample for every article, and that the technology was neither mature enough nor stable. Furthermore, if we were the only company in our sector that would introduce this technology, would we not run the risk of losing out against the competition? Were we not becoming our own worst enemy?

But the external application was only part of the equation. The product teams were even more used to their full physical sample set than our customers. For them, moving away from this and stepping out of the comfort zone would mean a major change. Hence, the external risks were often quoted as a showstopper for further expansion.

This was when we decided to make our customers part of the plan. If we could work with a very demanding, more traditional account and sell our ranges to them using virtual samples, would this not prove that the technology was indeed working and that moving out of the comfort zone was possible for everyone?

So we set out to work with a few major accounts and conducted the sell-in process using the new technology alongside a reduced set of physical samples. The new process was extremely well-received, and some of the accounts even went to so far as to request the technology for their internal purposes.

Furthermore, when speaking to the teams, we wanted to find out what their buying behaviors looked like. Did they not purchase products for their own personal use via the Internet? Was this not product that they had never physically seen and yet they were willing to spend their own money on it?

We also wanted to demonstrate to them that this change would not just mean that something was taken away from them, but they would get a great new way of working instead: something that could make their lives a lot easier.

4.2 *The Power of the Virtual Network*

The seasonal range had been defined, and our sourcing partners were busy creating the virtual samples. We had established a small team in HQ in order to refine the files upon arrival that would then be dispatched via Web to our showrooms in the pilot region. But how would the sales forces react? Were they ready for this major change in their day-to-day work? Could they cope with a very different, new way of conducting their business, and what if the retailers would not buy off virtual samples? What if they took their business elsewhere? After all, the aforementioned sell-in with major accounts was quite a different setup in a different region with customers that were very closely linked to the brand.

So, in order to alleviate some of these doubts and ensure that the sales force was well-prepared for the launch date, we put together an extensive training schedule and sent HQ teams into the region to train the sales force not only on the new tool, but also on the process and behavioral shift that was required to make this approach work.

It was important in this context that we provide the trainings in the local language and not assume that our company language, English, would be sufficient to drive such a game-changing project in the region.

4.3 Lessons Learnt

When we subsequently launched the season with the new process, it went rather smoothly, but there were a few bumps in the road as well. We encountered issues with the file quality from some of the factories, rendering rework in HQ higher than originally planned. File distribution also posed a problem due to file size and the fact that not all locations were equipped with the right bandwidth. The mixture of physical and virtual samples made the sell-in process somewhat more complex, as the sales responsible had to work both with a system and the actual sample. All in all, however, the first season already proved to be a major success with orders actually increasing, and no negative impact felt from the lack of physical samples.

4.4 Top Management Support: Make or Break

It has to be noted that the best-organized project, the most impressive innovation, and the most efficient processes and systems cannot be implemented successfully, if one key element is missing—top management support. This is where we were very lucky with the pilot region.

The regional head was extremely supportive of the project and committed himself and his team to making virtualization work. He removed many bumps in the road and proactively drove the change management process. He challenged but also supported his team, and therefore created a positive spirit and can-do attitude. He became an integral part of the virtual team and ensured that the vision was shared across the region. The pilot region spanned a huge geographic area with many different cultures, different economies, and different consumers—yet by taking ownership and committing to the change, the success of the project was made possible.

5 Expanding Virtualization from an Efficiency Driver to a Consumer Stimulus

The virtualization project could be best compared to growing a plant. You seed the idea; then you have to take care of it, water it, groom it, let it grow, and when it starts being a certain size, you ensure that it has room to expand and that it can stand for itself.

5.1 Overcoming Resistance to Sell-In

Following the success of the pilot project, senior management asked for more. A steering committee had been formed which had to decide on the next steps. We developed a rollout plan that not only covered the sell-in process, but would also consider the other application areas of virtualization, namely, cataloguing, e-commerce, brand marketing, customization, events, and ultimately design. We wanted to make our vision a reality—virtualization would be a new, revolutionary way of working. We would revolutionize the value chain. We would use one file in many different application areas and therefore achieve a myriad of benefits.

A first step, however, was to identify the future rollout for the sell-in process beyond the pilot region. Although customers and the management team in the pilot region had been very supportive and embraced the concept of virtualization, other markets were more demanding and doubtful with regards to the actual feasibility of changing the sell-in process and resistance was a lot higher.

In order to tackle this resistance, we decided that we would provide a certain safety net of physical samples for the first season that could be used as a fall-back solution in case the new technology was not accepted by retailers. Having all categories on board proved to be another important factor, which, however, put a burden on the factories producing the virtual samples as their workload increased dramatically.

A perfect showroom setup, well-functioning technology, and sound trainings prior to launch were paramount to further expand virtualization into other regions and with the rest of the customer base.

The sales forces needed to be shown what was in it for them, as just taking away their physical sample set and replacing it with a new technology that they would need to get used to was not a convincing proposition. Therefore, the ease of use of the system, reduction in setup time for the showrooms, and advantages that the retailers would see in the process, all had to be sold to the internal teams.

5.2 Cataloguing

In the meantime, the HQ teams were working on the other application areas that virtualization holds in store. Some of these areas were straightforward and could almost be treated as derivatives of the existing process. As an example, it is no longer necessary to order physical samples and conduct expensive photo shoots in order to create a catalogue. Instead, the already existing virtual samples can be used for the catalogues, thus making the process a lot less costly and also faster.

5.3 Customization

Consumers want brands to be relevant to them. One way of becoming relevant is making the consumer part of the product creation and letting them customize their very own product. Previously, the creation of these base files had to be outsourced to a third party, a process that was time consuming and complex. With virtualization, this process could be brought in-house, which allowed us to offer more models for customization.

5.4 *Brand Marketing*

Brand marketing campaigns often require product images. In the past, photo shoots would have to be conducted for every single product that we wanted to feature in a campaign. This is where our collaboration with the leader in 3D virtualization in the automotive industry helped us tremendously. After all, most images of cars these days are rendered products and not photographs. Using this experience and applying it to our industry was a key step forward.

5.5 *Design*

However, one important facet of the value chain has not been covered so far—design. The file creation to date is outsourced, which means that virtual files are earliest introduced to the value chain at the prototyping stage. The challenge is that designers create products without having all necessary information on hand, such as profitability drivers and engineering requirements. The current virtual technology is still rather technical, which creates immediate resistance with the design base. The fear of not being able to work creatively, being squeezed into a system, and having to work in a very technical way has so far hindered us from starting virtualization with design.

But technologies and process are being developed in close cooperation with design to create a design-friendly solution that is not only linked to the PLM systems and therefore allows price, materials, and engineering information, but also fosters creativity. This closes the loop in the value chain and makes design the driver of innovation.

5.6 *A New Consumer Experience*

It does not stop there. The opportunities that virtualization provide are unlimited. What started as an efficiency driver has turned into a consumer stimulus. By applying virtualization also in the direct interaction with the consumer, we can turn the consumer to a *prosumer*—an actively involved brand ambassador. Imagine having your own personal gym trainer at home—of course, virtually; creating your own virtual gym experience; sharing your product choices with your friends in social media; designing your own product and have your social network vote on your ideas. For our brand, this holds many opportunities as well. We can involve the consumers actively with the brand and make them part of the brand.

6 Making Innovation Sustainable

Innovation cannot be a one-off event. It is important to continuously experiment, to expand our horizons, and to look for the next step. We have to ensure that what we achieve in innovation gets grounded and builds the basis for further ideas.

In today's world, it is paramount to continuously innovate but also to make innovation sustainable. Making innovation sustainable means not just creating something from scratch and coming up with something completely new every time. Making innovation sustainable also means taking something existing and placing it into a different context.

Virtualization also fosters sustainability in a different context. It can enable companies to positively influence their environmental strategy. If we consider the amount of energy spent on producing samples and flying teams around the world to conduct product approvals, then it becomes very obvious that virtualization will be the future of the value chain in a greener, more ecological society.

7 Outlook

Virtualization is here to stay. It is a true game changer in the sporting goods and consumer goods industry. Our society is changing fast. People are online and accessible 24/7, consumers are more demanding than ever, and we have become more and more global. Our value chain now needs to follow suit.

What started out as "blue sky thinking" has become a game changer in the industry. The question of "what if we were to do things dramatically differently and what would it take" has been answered through the creation of virtualization. It has led to a paradigm shift with external partners, it has shown that looking beyond our own area of expertise can be a key enabler for the innovation process, and it has taught us a lot about change management and how virtual teams can turn a vision into reality.

Continuously improving virtualization and exploring and finding new application areas define the future of a very different value chain, a value chain that is faster, less costly, and highly innovative. To achieve this, it is paramount to go beyond what we know and foster collaboration networks with external partners, to learn from each other, apply best practices, and create innovation in places that formerly did not lend themselves to change.

Virtualization has proven that innovation is not limited to products. Innovation is a mindset. Innovation must be part of our culture. Innovation means openness, curiosity, and the courage to do things dramatically differently—against all odds.

Chapter 11
Technology Foresight: The Evolution of the Shell GameChanger Technology Futures Program

Roger Dennis, Tim Jones, and Leo Roodhart

Abstract Long-term planning is becoming increasingly problematic for organizations. Technology developments are taking pace at a speed that makes it difficult to leverage traditional linear forecasting methods. Shell is one of the most advanced users of techniques that are designed to explore possible alternate future scenarios. This article outlines why Shell developed a program called Technology Futures to look at areas outside of energy for insights and innovations that provided new opportunities for its business. In addition, the article details the next iteration of the program called Future Agenda, which is the world's largest foresight program, and the only one developed on an open source framework. Linking foresight to innovation in order to explore opportunities is an effective method of developing new high margin offerings.

1 The Challenges of Robust Foresight: Why Look Across Adjacent Sectors?

The silence was deafening. The chief scientist of the world's largest agri-tech company had just casually revealed that his company had developed a variety of corn that could sweat oil. For a while, nobody spoke. Seizing the moment, he went on to elaborate that they were planting it in such a way that meant the oil would be pumped out of the fields directly into storage and processing facilities. There was more silence.

R. Dennis (✉)
Growth Agenda Limited, 84 Brook Street, W1K 5EH, London
e-mail: now@rogerdennis.com

T. Jones
Growth Agenda Limited, London, UK

L. Roodhart
Said Business School, University of Oxford, Oxford, UK

S.P. MacGregor and T. Carleton (eds.), *Sustaining Innovation: Collaboration Models for a Complex World*, Innovation, Technology, and Knowledge Management,
DOI 10.1007/978-1-4614-2077-4_11,

Silence was not something that this group of people was accustomed to. As some of the most forward-thinking minds in Shell,[1] just a few minutes earlier, they were quite confident that they knew enough about biofuel developments to be able to sleep well at night. From the look on their faces, now sleep might not be so forthcoming that night.

It was 2004, and in addition to agriculture scientists, the small group of 40 people in the room included experts in online finance, robotics, aerospace, architecture, fast-moving consumer goods, and a plethora of other areas. With the exception of the hosts from Shell, none of them knew much about fossil fuels, and that was the reason they were there.

The Shell Technology Futures program had brought them together for 3 days to get a glimpse of how the next 20 years looked through their eyes. This group was assembled in Texas in the USA, but other groups had also convened in the EU with the same intention. The premise of the program was straightforward: to explore unfamiliar territory to uncover both risks and opportunities for the company.

Most foresight programs undertaken by organizations usually stay in familiar territory to develop their own particular view of the future. In doing so, they run the risk of being blindsided by developments outside their core business. They also miss the opportunity to proactively investigate potential high margin opportunities.

Furthermore, many strategy teams inside large organizations fail to fully comprehend that the most disruptive changes do not originate from the core of their sector, but from the edges when disciplines spark off each other.

Perhaps the most well-known example of this is how Apple has come to dominate the music industry. In 2000, prior to the introduction of the iPod, it is safe to say that music industry executives would have paid little notice to the potential impact that a computer manufacturer could have on their business. In doing so, they failed to see how digitalization of a business model leaves it open to disruption from the fringes of their industry.

At Shell, Technology Futures was an initiative that specifically addressed these issues by seeking insights from specialists outside the core business. The mix of experts assembled was chosen because their backgrounds overlapped at a sweet spot: the intersection between new technologies, societal trends, and consumer behaviors.

Each specialist was selected through an intensive research process that first identified a technology that had the potential to either disrupt the core business or provide an opportunity. This selection required the involvement of technology watchers, commentators, and analysts to supply the depth of knowledge necessary to narrow the extensive list down to a handful of sectors and industries.

Following this process, a shortlist was drawn up of the most promising thought leaders in each of the target fields. The criterion for selection was not only limited solely to one profession within that field, but also included academics, entrepreneurs,

[1] Shell is one of the world's largest companies and has the tenth largest market value in the world (according to *Forbes* in April 2011). It is an energy company with most revenue from oil and gas and also has a large renewable energy programme.

specialist journalists, and commentators. Structured interviews were undertaken with each candidate before the final list was drawn up. Those that made the cut were then invited to spend 3 days in immersive debate and discussion at various locations around the world.

The final mix of attendees included experts from the area of architecture, information technology, technology ethnography, urban planning, synthetic biology, space robotics, and superconductivity. Organizations represented were equally diverse—ranging from Victoria University in New Zealand to Nokia in Tokyo. This diversity was critical in order to introduce insights from adjacent sectors that were outside of the normal areas that Shell would scan.

The focus for each of the sessions was to look 20 years out to identify drivers and impacts not only in their respective specialties, but also to identify cross-over points between fields that might create entirely new opportunities. Careful and expert facilitation was then required to separate probabilities from possibilities.

Each session was operated under Chatham House rules: "When a meeting, or part thereof, is held under the Chatham House Rule, participants are free to use the information received, but neither the identity nor the affiliation of the speaker(s), nor that of any other participant, may be revealed." This set the environment for attendees to have a much more open discussion than they would normally consider and share insights that would normally remain behind closed doors. There were no competing organizations present, and one of the foundations of the sessions was that attendees would take away from the sessions as much as they contributed.

The most critical aspect of any program like this is the return on the investment, and this hinges on how the output is framed. By identifying potential opportunities and positioning them on a timeline, Technology Futures made it possible to make well-informed bets on areas that warranted further investigation. In addition, the program also identified likely technology pathways and technology-enabled impacts and did so in a format suitable for linking into existing scouting, scenario, and business strategy processes.

2 Linking the Long and Short Term: How Technology Futures Was Used In-House?

Building physical and conceptual towers to look further out into the horizon has been a well-established approach in Shell since the late 1960s. Scenario planning—preparing for possible future contexts—has been a preferred route to consider how uncertainties might impact the current context. The Black Swan[2] (Taleb 2010) cannot be predicted: we do not know what we do not know. However, *what-if?* questions can be formulated within a certain context/scenario so that one is prepared if the Black Swan hits the windshield.

[2]The Black Swan theory deals with events that are a surprise (to the observer) and have a major impact. After the fact, the event seems perfectly rational in hindsight.

Table 11.1 The Shell GameChanger team was tasked with discovering the unknown unknowns. Other domains of knowledge were already allocated to various groups within the company

Unknown	Things we do not know we know (*Knowledge sharing*)	Things we do not know we do not know (*GameChanger*)
Known	Things we know we know (*Best practice*)	Things we know we do not know (*Research and Development*)
	Knowns	Unknowns

Scenario planning produces a small set of possible contexts that a firm may be in, contexts that are unfolding beyond the control of that firm. The methodology is especially valuable for activities that have long-lead times for technological innovation, and capital expenditures that are financed over decades, such as those in the oil industry.

However, a shift happened in the last decade.

In an increasingly turbulent world, "old-fashioned" long-term planning based on forecasts becomes impossible. Uncertainty rules and the turbulence of our business environment reaches a new climax year after year. The end of strategic planning is in sight.

The main culprit and accelerator here is technology development. Who would have thought that networking tools, developed to amuse teenagers, such as Twitter and Facebook, would be able to overthrow governments?

It has been said before: the Stone Age did not end due to a lack of stones. One technology break-through in battery technology or superconductivity might completely change the energy world and, with it, the massive oil and gas companies that now fuel the world economy.

So if strategic planning process no longer works, then what?

Global companies, to survive now, need another skill apart from being prepared for major unexpected events with their scenarios: they need the right, leading-edge technologies to become available the moment they are required. As new technology takes several decades to develop, this means that technology and long-term scenarios have to relate in unprecedented ways.

Scenario planners and strategists deal with the strategic options that possible future contexts may close down or open up. On another spectrum, technologists and scientists are monitoring how whole families of technology options (such as GMOs, nano-technologies, super-conductivity, bio-engineering, etc.) may convert current impossibilities into future possibilities.

To this end, when the Technology Futures program was started as part of Shell's strategic innovation effort in 2003, the underlying philosophy was based on known information. Shell Research was already tasked with understanding what Shell did not know and an in-house innovation program called GameChanger addressed information that the company did not know that it knew, as presented in Table 11.1. What remained was the knowledge completely outside Shell's core business, outside its scientists' research portfolios and outside of its unmined internal knowledge.

Essentially, what was left was the knowledge that Shell did not know that it did not know.

To address this, a team of scouts was established to scan the world of technology outside the company: talk to unknown people in the unknown Silicon Valleys of this world, find the unknown technology threats, and identify unheard of opportunities.

The Technology Futures workshops were organized in support of this search. In each workshop, the assembled thought leaders asked which technology breakthrough they saw coming in their industry in the next 20 years. This generated technology pathways for different technology areas, such as medicine, IT, communication, agriculture, and more. Sometimes, two pathways would cross, indicating a possible disruption. These crossovers were discussed and further documented. The end result was a technology map for the next 20 years.

So how were these pathways used in the scenario process?

Within the contextual possibilities scenarios offer, various possible futures of a company can be identified. With two or three purpose-built scenarios, it is possible to develop a handful of possible future portfolios for a company. These pathways then need to be built to obtain strategically dominant positions within each of these futures; that is, one needs to find out what one will need to have in place (both technology and business-wise) to start moving the firm in the new direction of the aspired portfolio via one or more paths. The pathways connect (aspired) Company Future to (actual) Company Now.

New and unknown technologies and new business models based on new technologies shape each of these pathways. It is in these very pathways where technologists and strategists must coproduce new plausibilities in a ground neither of them owns alone, and this was the essence of the Technology Futures program.

3 An Example of a Power Company

To illustrate how this works, consider an artificial example of a cheap, coal-burning power company. Let us assume that two scenarios have been developed:

(a) An "open doors" scenario where the world would have further globalized, free markets rule, and global warming is seen as a global issue that can be solved together.
(b) A "closed doors" scenario where the world would have become much more confined, regional power structures have formed, and no global agreements can be reached on anything.

The current pathway, as a trend projection, would keep the company in the future producing cheap energy by the continued burning of (hydro)-carbons, as it is doing now. However, if the company considered the first scenario as unfolding, in which not becoming green becomes prohibitive, a first possible company future would entail becoming a green energy company that burns carbon cleanly.

If it instead considered that the world might unfold as per the second, closed-door scenario, a good option for it would be to enter new energy systems early. That option, through first mover advantages in its protected regions, would allow it to

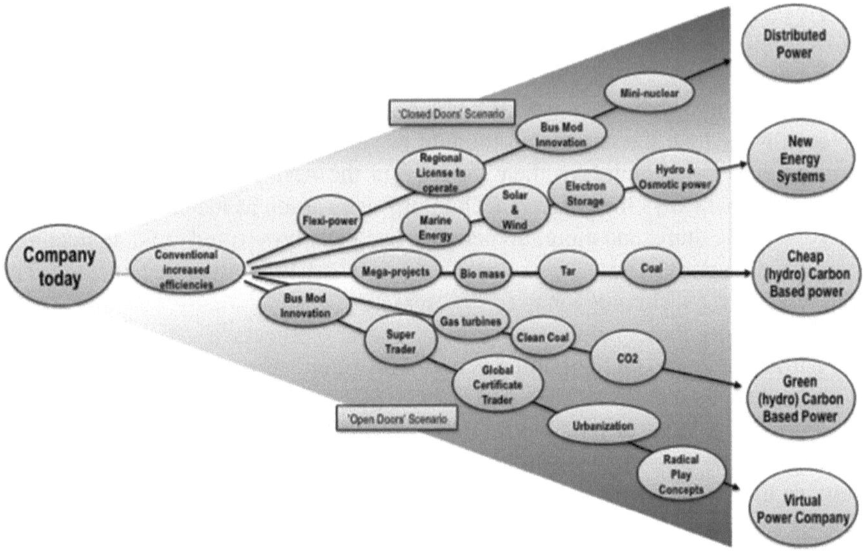

Fig. 11.1 Five possible future configurations for a power company

dominate market segments there. These two futures are relatively small deviations from the current strategy of being a cheap (hydro-)carbon-burning power producer.

Two more extreme future possibilities are at the edges of what is considered plausible today.

In the "open doors" scenario, it is possible to imagine a radically different business with the company becoming a virtual power player with no self-owned assets like power plants. In developing such a future for itself, the company might become a global trader in electrons, green certificates, and carbon credits; making money on a unique ability to optimize global energy flows and store electrons when they are cheap and sell when expensive.

In the "closed door" scenario, the firm may instead determine to pursue another radical pathway, where it would abandon large-scale power generation and focus on distributed energy; with small combined heat and power units for individuals, or industrial parks or city scale operations where it can again profit form first mover advantages.

The five possible future configurations for the company, all illustrated in Fig. 11.1, are plausible and realistic, but each has a distinct, unique development path. For example, in pursuing its path toward becoming a green (hydro-) carbon-based power company, it would need to invest in and master gas turbines, clean coal, and CO_2 capture and storage activities.

If instead the company wanted to transform itself into a new energy systems player, it would have to master marine, solar, wind, hydro, and osmotic power, as well as electron storage technologies and business models. Insights obtained by developing the Technology Futures would fill the pathways with technologies previously unknown to the power company developed for a completely different industry.

During this exercise, technologists must leave the comfort of their own specific deep ("vertical") expertise and relate it to similar expertise in other fields based on other projects in their domain. This is not as wide as considering the whole corporate set of options, but wider than their everyday thinking. In the same way, strategists must obtain deeper understanding of what different technologies offer (and cannot deliver) so that they can compare apples and oranges together and make sense of how to make alternative pathways to the future develop best. The "so what" significance of these activities relate the pathways they articulate to the fit of the strategy and with the overall contextual future scenario that is under consideration.

4 Shell's GameChanger Methodology

Linking long-term global scenarios to technology futures thus connect two forms of innovation—the top-down perspective and the bottom-up one that Hamel's idea of "bringing Silicon Valley inside" large companies famously espoused in the 1990s (Hamel 1999).[3] It connects both possibilities in an ecosystem in which both can live, flourish, and help each other.

The content of the new and specific projects on the pathways can come from a variety of sources; scientists, developers, customers, partners, or users.

Any company, however, would benefit from a structured innovation-generating and assessment methodology. The methodology used within Shell is called GameChanger, which funnels ideas from internal and external sources and pushes these ideas through a series of assessment teams. The first group is the GameChanger team, plus a number of peers, followed by a senior panel, and ending with the executive committee, until only the technically and financially feasible ideas are left on the table.

After emerging from the GameChanger funnel, approved ideas are converted into action projects. These projects can range from in-house research and development to joint ventures with outside companies. In-house R&D results in new technology for the company. External partnerships and spinouts create new businesses in which the domain can be explored and the appropriate skills-sets can be acquired. Projects that do not fit or spawn a domain are quickly abandoned.

In summary, linking long-term scenarios to technology futures offers a rigorous approach for connecting long-term aspirations and short-term projects, strengthening the otherwise weak and mixed middle.

---

[3]Gary Hamel is a leading management thinker who has published a number of highly regarded books. In 2008, the *Wall Street Journal* ranked him as one of the world's most influential business thinkers.

Fig. 11.2 Screenshot of the Future Agenda Web site

5 Beyond Technology Futures: Future Agenda

Arguably, the thought leaders in foresight are Shell and IBM, and with the Technology Futures approach echoed in the IBM GIO[4] initiative, clearly the method of such programs has considerable merit.

However, just as both programs evolved from a need to stay abreast of fast-moving technologies, how would such a program itself evolve?

This was the question that triggered the birth of Future Agenda—the world's largest foresight initiative (see Fig. 11.2 below for a view of the online resource). The essence of the initiative boiled down to a thought experiment focusing on one of the insights from Technology Futures: What would happen if you took one of the dominant changes in computing—open source software—and applied the philosophy to the field of foresight?

This drove the development of Future Agenda.

Sponsored by Vodafone Group, the discussion began by identifying 16 issues that were viewed as the most important for the next decade, such as water, connectivity, choice, and migration. A number of experts from academia and industry were then invited to offer an initial view on these issues.

These expert-led perspectives were then distributed online and used in the next stage of the program to stimulate debate and discussion in workshops around the world. In total, Future Agenda held 50 workshops in 25 locations around the world gathering insights from 2000 people. The audiences included sector experts, entrepreneurs, government ministers, scientists, CEOs, and innovators from many countries, including Australia, New Zealand, China, India, Germany, Spain, the UK, and the USA.

---

[4] The IBM Global Innovation Outlook (GIO) started in 2004 and echoed the same approach as the Shell Technology Futures programme. IBM bought together experts from around the world to address big issues facing the world. The aim of the programme was to identify new business opportunities.

In contrast to closed corporate foresight programs, all the discussion from the workshops was posted online as the program developed. In this manner, people who had not been able to attend the events could give feedback and comment on the material in the same way as those who had attended. Online posting of the material also allowed participating organizations to immediately feedback some of the material into their own strategic processes.

From the start, the original intention for running the Future Agenda program was not just to create a broader, deeper, and richer view of the future. It was also to provide this view openly for all to use. To further this, a key part of the design of the program was the wide dissemination of the output. As such, the output was provided in a number of forms, including a book, presentations, downloadable documents, and a Web site with linkages and data (http://www.futureagenda.org). On top of that, a card-based version of the insights was designed for use in workshops, and in some cases, iPhone and iPad apps were created and shared.

6 In Practice: Using the Insights from Future Agenda

As sponsor of the program, Vodafone Group has clearly used the insights for a wide range of activities, but other companies and organizations have also been making use of the insights for a wide range of purposes. These organizations can be split into three core groups:

First, organizations familiar with foresight have been implementing the Future Agenda program outputs into their ongoing activities. Shell, IBM and the Government of Singapore are just some who have used the books, presentations, and Web site to share the insights across parts of their organizations and integrate them within current scenario development and foresight activities. Although only a few of the insights from the Future Agenda program are totally new to these organizations, many have found the perspectives have complemented existing views and others have added different takes.

Second, other companies have moved straight into using the Future Agenda insights as stimulus for innovation—either identifying new spaces to potentially explore and develop totally new products, or as input to incremental brand and category development activities.

The likes of PepsiCo, Mars, Discovery Channel, Abbott, and Castrol were some of the first to do this during the early stages of the program, but as the insights were more widely shared, more and more companies across the USA, the EU, and Asia have also joined in. Although at the time of writing, none of the corresponding new innovations have hit the market, we can see a wide range of products, services, and businesses moving through varied development pipelines.

What is significant about many of these innovations is that they are inherently cross-disciplinary in nature. They are taking insights from one region or sector and using them to stimulate new thinking within another area. They are mixing outside-in and inside-out approaches to identify and detail new opportunities beyond the usual portfolio.

A third group of organizations have used the Future Agenda insights less for immediate innovation activities, but instead focused on challenging assumptions and broadening thinking outside the usual core sectors or time horizons. Companies, such as Microsoft, Reckitt Benckiser, and IBM, have used some of the insights within their businesses as catalysts for wider debates on emerging challenges and opportunities and how to address them.

While Vodafone Group has also done most of the above efforts in different parts of the world, additional approaches have also been undertaken. In the UK and the USA, Vodafone Global Enterprise has used the Future Agenda insights as a core element within collaborative innovation events with some of its major corporate customers. For example, in Brussels, a series of debates around some of the core issues, such as trust, privacy, identity, and choice, have been used to engage the EU in new ways, and at the Group level, insights from the program have been put through a mobility lens to create a more sector-focused view on the future of mobile over the next decade.

However, perhaps most notably, Vodafone Turkey has embraced the Future Agenda program as a platform for stimulating wider discussion and debate on the future of Turkey itself. As well as hosting the global launch and engaging a broad range of stakeholders from the start in Turkey, additional insights on local perspectives were developed through discussion with key business and NGO leaders. In addition, CNN has run a series of programs sharing the insights across a broad platform; other news and print media have made extensive use of the material to discuss pivotal issues, and a number of debates within universities on key topics have also occurred.

Given that Turkey is such a large, high growth economy with a historical position at the East–West frontier, it is notable how the Future Agenda platform has resonated so well as a catalyst for discussion on where the country could be in the future. Championed personally by Vodafone Turkey's CEO who made herself widely available to the media to inform and discuss the program, other countries and regions are now also using the material for similar purposes.

By its very nature, the Future Agenda program engaged new communities as part of its insight-gathering phase and is increasingly doing the same now that these insights are being shared. New audiences for foresight beyond the usual suspects have been found, new uses of the associated insights are coming to the fore and new propositions are being suggested. Moreover, given the open nature of the program, the team running it only really knows about the uses that they have either been involved with or directly notified of. Based on Google Analytics data for the core Web site, it is clear that many others are also making use of the program output for their own purposes.

7 Open Foresight as a Core Asset

So what is the future of open foresight? How will it be used, and what will be the main advantages that it provides to organizations, be they government or corporate?

To start with, we need to recognize that there is a substantial community of people and organizations that see little or no value in open foresight. Some of these concerns come from the heart of the futures' community and clearly see this program as a threat to their businesses. From a principled perspective, there is little that can really be said about this, other than open foresight creates a wider audience and opportunities for interpretation.

It is worth highlighting here the associated risk of sticking to the recognized experts and not engaging with the naïve emerging voices. At the start of the Future Agenda program, one notable leader at a renowned academic institution questioned why go to the effort of running numerous workshops around the world when the best brains could be found in his university. He said, "If you need to know what will be important in the future, just ask us."

Others see that there is inherent risk in relying on the wisdom of crowds and the "Tower of Babel." How can mass engagement on foresight guarantee the quality and rigor that can be provided by those better trained in the art of foresight? While these concerns are all clearly valid, others would say that one of the biggest problems with today's foresight methods is that they are focused too much on a narrow community who engage repeatedly with the usual suspects.

As Shell's Technology Futures programs, IBM's GIO, and the Future Agenda program have all demonstrated, engaging a wider range of views in a coherent manner can highlight useful left-field views that can take organizations into new areas that traditional scenarios or futures programs have not. Clearly, the noise from white space foresight where anyone and everyone can put in their pennyworth has significant risks that technology cannot yet overcome. For now, however, if focused on specific topics, stimulated but not bound by pre-research and orchestrated in a manner that encourages debate and dialogue, then open foresight programs can clearly deliver additional and complementary views to the futures' mainstream views.

The above notwithstanding, it is clear to us that open foresight does have a place within the futures' repertoire. It is not a replacement for other established approaches, but provides a complementary means of gaining access to a wider range of views. While much of the core outputs of an open foresight program may overlap with traditional approaches, there are four areas where we have seen distinctive benefit:

7.1 Sharing Views

Although this is clearly an open program, contributions to the content of Future Agenda are confidential and anonymous. This allows organizations to share perspectives that they would normally hold close. By providing a platform for the open sharing of multiple views, companies and governments can more readily contribute their thoughts and see how others react to them.

7.2 *Engaging a Broader Church*

The very nature of open foresight programs allows for a wider audience to be engaged. Irrespective of the quality of insights gained, some would argue that there is benefit to be gained from stimulating a larger section of society to think about the future. Interaction with a broader church provides opportunities for new connections to be made outside the norm and hence outcomes may well be different than in usual foresight.

7.3 *More Weak Signals*

By engaging a wider, global audience, more views are brought into the mix from more countries and a wide range of experiences, ages, and backgrounds. As such, the broader mix inherently means more data, and within this data, there are many more smaller views than traditional approaches uncover. If these views are identified as additional weak signals that sit alongside the high probability foresight, and used as such, they provide a useful additional layer.

7.4 *Stimulating Higher Levels of Innovation*

Through accessing a wider audience on a broader range of issues, many see that open foresight approaches provide greater opportunity for the clashing of views, interaction between sectors, stimulus of new ideas, and as a consequence, the potential for higher levels of innovation.

8 Conclusion

In a study of 19 multinational companies, Gemunden and Rohrbeck highlighted that the "…practice of linking people to pass on foresight insights to the ones that can start new innovation initiatives has also proven to be an effective method to enhance the innovation capacity of a firm" (Rohrbeck and Gemunden 2011).

The Technology Futures program did exactly that for Shell in several areas. For example, the program drew attention to the potential of marine algae as a production source for biofuels. As a result in 2007, it commenced a joint venture in Hawaii with HR Biopetroleum and made a considerable investment in the construction of a demonstration facility to grow marine algae and produce vegetable oil for conversion into biofuel.

As the pace of technology changes increases, the standard strategic planning approach leaves organizations vulnerable to risks and unable to take advantage of opportunities. The best not only understand the linkages between foresight, strategy, and innovation, they also position themselves to leverage the knowledge that is developed from this understanding.

Organizations no longer need deep pockets and global reach to run their equivalent of the Shell Technology Futures program or the IBM Global Innovation Outlook initiative. Open source futures programs, such as Future Agenda provide the insights.

However, while this removes some barriers, this knowledge alone is not enough to provide advantage. To fully leverage an open source foresight program requires a sharp strategy team that can direct a strong innovation group toward opportunities that are ripe for disruption.

And that is where the fun begins.

9 Resources

A PDF version of the book from Shell's 2007 Technology Futures program can be downloaded from:

http://www-static.shell.com/static/innovation/downloads/innovation/technology_futures.pdf.

The Future Agenda Web site is at http://www.futureagenda.org, and the eBook can be downloaded from http://www.futureagenda.org/pg/cx/view/450?view=acs.

References

Hamel, G (1999) Bringing Silicon Valley inside. *Harv Bus Rev* 77:70–84, 183

Rohrbeck R, Gemunden H G (2011) Corporate foresight: Its three roles in enhancing the innovation capacity of a firm. *Technol Forecast Soc* 78:231–243

Taleb N (2010) *The black swan: The impact of the highly improbable*. Random House, New York

Index

S.P. MacGregor and T. Carleton (eds.), *Sustaining Innovation: Collaboration Models for a Complex World*, Innovation, Technology, and Knowledge Management,
DOI 10.1007/978-1-4614-2077-4, © Springer Science+Business Media, LLC 2012

MIX
Papier aus verantwortungsvollen Quellen
Paper from responsible sources
FSC® C105338

If you have any concerns about our products,
you can contact us on
ProductSafety@springernature.com

In case Publisher is established outside the EU,
the EU authorized representative is:
Springer Nature Customer Service Center GmbH
Europaplatz 3, 69115 Heidelberg, Germany

Printed by Libri Plureos GmbH
in Hamburg, Germany